Universitext

Springer

New York
Berlin
Heidelberg
Barcelona
Hong Kong
London
Milan
Paris
Singapore
Tokyo

Universitext

Editors (North America): S. Axler, F.W. Gehring, and K.A. Ribet

Aksoy/Khamsi: Nonstandard Methods in Fixed Point Theory
Andersson: Topics in Complex Analysis
Aupetit: A Primer on Spectral Theory
Bachman/Narici/Beckenstein: Fourier and Wavelet Analysis
Balakrishnan/Ranganathan: A Textbook of Graph Theory
Balser: Formal Power Series and Linear Systems of Meromorphic Ordinary
  Differential Equations
Bapat: Linear Algebra and Linear Models (2nd ed.)
Berberian: Fundamentals of Real Analysis
Booss/Bleecker: Topology and Analysis
Borkar: Probability Theory: An Advanced Course
Böttcher/Silbermann: Introduction to Large Truncated Toeplitz Matrices
Carleson/Gamelin: Complex Dynamics
Cecil: Lie Sphere Geometry: With Applications to Submanifolds
Chae: Lebesgue Integration (2nd ed.)
Charlap: Bieberbach Groups and Flat Manifolds
Chern: Complex Manifolds Without Potential Theory
Cohn: A Classical Invitation to Algebraic Numbers and Class Fields
Curtis: Abstract Linear Algebra
Curtis: Matrix Groups
DiBenedetto: Degenerate Parabolic Equations
Dimca: Singularities and Topology of Hypersurfaces
Edwards: A Formal Background to Mathematics I a/b
Edwards: A Formal Background to Mathematics II a/b
Foulds: Graph Theory Applications
Friedman: Algebraic Surfaces and Holomorphic Vector Bundles
Fuhrmann: A Polynomial Approach to Linear Algebra
Gardiner: A First Course in Group Theory
Gårding/Tambour: Algebra for Computer Science
Goldblatt: Orthogonality and Spacetime Geometry
Gustafson/Rao: Numerical Range: The Field of Values of Linear Operators
  and Matrices
Hahn: Quadratic Algebras, Clifford Algebras, and Arithmetic Witt Groups
Holmgren: A First Course in Discrete Dynamical Systems
Howe/Tan: Non-Abelian Harmonic Analysis: Applications of $SL(2, R)$
Howes: Modern Analysis and Topology
Hsieh/Sibuya: Basic Theory of Ordinary Differential Equations
Humi/Miller: Second Course in Ordinary Differential Equations
Hurwitz/Kritikos: Lectures on Number Theory
Jennings: Modern Geometry with Applications
Jones/Morris/Pearson: Abstract Algebra and Famous Impossibilities
Kannan/Krueger: Advanced Analysis
Kelly/Matthews: The Non-Euclidean Hyperbolic Plane
Kostrikin: Introduction to Algebra
Luecking/Rubel: Complex Analysis: A Functional Analysis Approach
MacLane/Moerdijk: Sheaves in Geometry and Logic
Marcus: Number Fields
McCarthy: Introduction to Arithmetical Functions

(continued after index)

Peter Morris

Introduction to Game Theory

With 44 Illustrations

Springer

Peter Morris
Mathematics Department
Penn State University
University Park, PA 16802
USA

Editorial Board
(North America):

S. Axler
Mathematics Department
San Francisco State University
San Francisco, CA 94132
USA

F.W. Gehring
Mathematics Department
East Hall
University of Michigan
Ann Arbor, MI 48109
USA

K.A. Ribet
Mathematics Department
University of California at Berkeley
Berkeley, CA 94720-3840
USA

Mathematics Subject Classifications (1991): 90Dxx, 90C05

Library of Congress Cataloging-in-Publication Data
Morris, Peter, 1940–
    Introduction to game theory / Peter Morris.
       p. cm. — (Universitext)
    Includes bibliographical references and index.
    ISBN 0-387-94284-X
    1. Game theory.   I. Title.
QA269.M66   1994   94-6515
519.3—dc20

Printed on acid-free paper.

Production managed by Laura Carlson; manufacturing supervised by Gail Simon.
Camera-ready copy prepared by the author using *AMS*-LaTeX.
Printed and bound by R. R. Donnelley & Sons, Harrisonburg, VA.
Printed in the United States of America.

9 8 7 6 5 4 3

ISBN 0-387-94284-X Springer-Verlag New York Berlin Heidelberg
ISBN 3-540-94284-X Springer-Verlag Berlin Heidelberg New York   SPIN 10762476

To my mother, Opal Morris, and to
the memory of my father, W. D. Morris

Preface

The mathematical theory of games has as its purpose the analysis of a wide range of competitive situations. These include most of the recreations which people usually call "games" such as chess, poker, bridge, backgammon, baseball, and so forth, but also contests between companies, military forces, and nations. For the purposes of developing the theory, all these competitive situations are called *games*.

The analysis of games has two goals. First, there is the descriptive goal of understanding why the parties ("players") in competitive situations behave as they do. The second is the more practical goal of being able to advise the players of the game as to the best way to play. The first goal is especially relevant when the game is on a large scale, has many players, and has complicated rules. The economy and international politics are good examples.

In the ideal, the pursuit of the second goal would allow us to describe to each player a strategy which guarantees that he or she does as well as possible. As we shall see, this goal is too ambitious. In many games, the phrase "as well as possible" is hard to define. In other games, it can be defined and there is a clear-cut "solution" (that is, best way of playing). Often, however, the computation involved in solving the game is impossible to carry out. (This is true of chess, for example.) Even when the game cannot be solved, however, game theory can often help players by yielding hints about how to play better. For example, poker is too difficult to solve, but analysis of various forms of simplified poker has cast light on how to play the real thing. Computer programs to play chess and other games can be written by considering a restricted version of the game in which a player can only see ahead a small number of moves.

This book is intended as a text in a course in game theory at either the advanced undergraduate or graduate level. It is assumed that the students using it already know a little linear algebra and a little about finite probability theory. There are a few places where more advanced mathematics is needed in a proof. At these places, we have tried to make it clear that there is a gap, and point the reader who wishes to know more toward appropriate sources. The development of the subject is introductory in nature and there is no attempt whatsoever to be encyclopedic. Many interesting topics have been omitted, but it is hoped that what is here provides a foundation for further study. It is intended also that the subject be developed rigorously. The student is asked to understand some serious mathematical reasoning. There are many exercises which ask for proofs. It is also recognized, however, that this is an applied subject and so its computational aspects have not at all been ignored.

There were a few foreshadowings of game theory in the 1920's and 1930's in the research of von Neumann and Borel. Nevertheless, it is fair to say that the subject was born in 1944 with the publication of [vNM44]. The authors, John von Neumann and Oskar Morgenstern, were a mathematician and an economist, respectively. Their reason for writing that book was to analyze problems about how people behave in economic situations. In their words, these problems "have their origin in the attempts to find an exact description of the endeavor of the individual to obtain a maximum of utility, or, in the case of the entrepreneur, a maximum of profit."

Since 1944, many other very talented researchers have contributed a great deal to game theory. Some of their work is mentioned at appropriate places in this book.

In the years after its invention, game theory acquired a strange reputation among the general public. Many of the early researchers in the subject were supported in part or entirely by the U.S. Department of Defense. They worked on problems involving nuclear confrontation with the Soviet Union and wrote about these problems as if they were merely interesting complex games. The bloody realities of war were hardly mentioned. Thus, game theory was popularly identified with "war gaming" and was thought of as cold and inhuman. At the same time, the power of the theory was exaggerated. It was believed that it could solve problems which were, in fact, far too difficult for it (and for the computers of that time). Later, in reaction to this, there was a tendency to underestimate game theory. In truth, it is neither all-powerful nor a mere mathematician's toy without relevance to the real world. We hope that the usefulness of the theory will become apparent to the reader of this book.

It is a pleasure to thank some people who helped in moving this book from a vague idea to a reality. Mary Cahill, who is a writer of books of a different sort, was unfailing in her interest and encouragement. Linda Letawa Schobert read part of the first draft and offered many suggestions which greatly improved the clarity of Chapter I. Ethel Wheland read the entire manuscript and corrected an amazing number of mistakes, inconsistencies, and infelicities of style. She is a great editor but, of course, all the errors that remain are the author's. Finally, thanks are due to several classes of students in Math 486 at Penn State who used early versions of this book as texts. Their reactions to the material had much influence on the final version.

Contents

List of Figures

1
Games in Extensive Form

All the games that we consider in this book have certain things in common.
These are:

- There is a finite set of *players* (who may be people, groups of people, or more abstract entities like computer programs or "nature" or "the house").
- Each player has complete knowledge of the rules of the game.
- At different points in the game, each player has a range of choices or *moves*. This set of choices is finite.
- The game ends after a finite number of moves.
- After the game ends, each player receives a numerical *payoff*. This number may be negative, in which case it is interpreted as a loss of the absolute value of the number. For example, in a game like chess the payoff for winning might be +1, for losing −1, and for a draw 0.

In addition, the following are properties which a game may or may not have:

- There may be *chance moves*. In a card game, the dealing of the hands is such a chance move. In chess there are no chance moves.
- In some games, each player knows, at every point in the game, the entire previous history of the game. This is true of tic-tac-toe and backgammon but not of bridge (because the cards dealt to the other players are hidden). A game with this property is

said to be *of perfect information*. Note that a game of perfect information may have chance moves. Backgammon is an example of this because a die is rolled at points in the game.

We have just said that the players receive a *numerical* payoff at the end of the game. In real conflict situations, the payoff is often something non-quantitative like "happiness," "satisfaction," "prestige," or their opposites. In order to study games with such psychological payoffs, it is first necessary to replace these payoffs with numerical ones. For example, suppose that a player in a certain game can win one of three prizes:

- A week in Paris.
- A week in Hawaii.
- Eight hours in a dentist's chair.

Different people would assign different "happiness ratings" to these prizes. For a person with an interest in French culture, rating them as 100, 25, −100, respectively, might be reasonable. To a surfer, the ratings might be 10, 100, −100. The point is that we are assuming that this conversion of nonquantitative payoffs to numerical ones can always be done in a sensible manner (at least in the games we consider).

It is natural to represent a game by means of a "tree diagram." For example, consider the following simple game (called Matching Coins). There are two players (named Thelma and Louise); each conceals a coin (either a penny or a nickel) in her closed hand, without the other seeing it. They then open their hands and if the coins are the same, Thelma takes both of them. If they are different, Louise takes both. A tree diagram for Matching Coins is shown in Figure 1.1.

The little circles in the diagram are called vertices, and the directed line segments between them are called edges. A play of the game starts at the top vertex (labeled "Thelma") and arrives, via one of the two vertices labeled "Louise," at one of the four vertices at the bottom. Each of these is labeled with an ordered pair of numbers which represent the payoffs (in cents) to Thelma and Louise, respectively. For example, if Thelma holds a penny, and Louise a nickel, then the game moves from Thelma's vertex by way of her left-hand edge, and leaves Louise's left-hand vertex by way of her right-hand edge. The bottom vertex reached is labeled $(-1, 1)$. This means that Thelma loses +1 ("wins" −1) and that Louise wins +1.

In this chapter, we rigorously develop this idea of tree diagrams. We also extend it to games with chance moves, and introduce a definition of strategy.

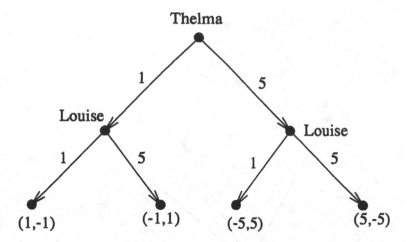

FIGURE 1.1. Matching Coins.

1.1. Trees

A *directed graph* is a finite set of points, called *vertices*, together with a set of directed line segments, called *edges*, between some pairs of distinct vertices. We can draw a directed graph on the blackboard or on a piece of paper by drawing small circles for the vertices and lines with arrowheads for the edges. An example is given in Figure 1.2. It has five vertices and five edges. The theory of graphs (both directed and not directed) is a large and interesting subject. See [BM76] for more information.

We will use uppercase letters like G or H or T to denote directed graphs. The vertices of such a graph will be denoted by lowercase letters like u or v. Subscripts will be used sometimes. Edges will be denoted by, for example, (u, v). Here, the edge goes *from u to v*. For a directed graph G, the set of all vertices is denoted $V(G)$.

A *path* from a vertex u to a vertex v in a directed graph G is a finite sequence $(v_0, v_1, \ldots, v_n)$ of vertices of G, where $n \geq 1$, $v_0 = u$, $v_n = v$, and (v_{i-1}, v_i) is an edge in G for $i = 1, 2, \ldots, n$. For example, in the directed graph of Figure 1.2, (e, c, a) and (e, c, b, e) are paths. A *tree* is a special kind of directed graph. We have the following:

DEFINITION 1.1. A directed graph T is a *tree* if it has a distinguished vertex r, called the *root*, such that r has no edges going into it and such that for every other vertex v of T there is a unique path from r to v.

The example in Figure 1.2 is clearly not a tree. A directed graph consisting of a single vertex and no edges is a (trivial) tree.

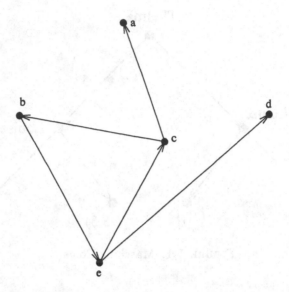

FIGURE 1.2. A directed graph.

Usually, but not always, we draw trees in such a way that the root is at the top. An example is given in Figure 1.3; all the vertices have been labeled. This tree has eleven vertices and ten edges.

Now let T be a tree. We make a few definitions. A vertex v is a *child* of a vertex u if (u, v) is an edge. Also, in this case, u is the *parent* of v. In the tree of Figure 1.3, vertex f is a child of b, vertex g is a child of c, a is the parent of both d and e, and the root is the parent of a, b, and c. In any tree, the set of children of u is denoted $\mathrm{Ch}(u)$. Notice that a vertex may have many children. The root is the only vertex without a parent. A vertex without any children is called *terminal*. In our example, the terminal vertices are d, e, f, i, j, and h. A vertex which is not the root and not terminal is called *intermediate*. In our example, the intermediate vertices are a, b, c, and g. A vertex v is a *descendant* of a vertex u if there is a path from u to v. In this case, u is an *ancestor* of v. Thus the root is an ancestor of every other vertex. In the example, c is an ancestor of j, and i is a descendant of c.

The *length* of a path is the number of edges in it. Paths in trees do not cross themselves; that is, they consist of distinct vertices. This follows from Exercise (8). Thus, a path in a tree has length at most the number of vertices minus one. The *depth* of a tree is the length of the longest path in it. In Figure 1.3, (c, g, i) is a path of length 2, as is (root, b, f). It is clear that the longest path in a tree starts at the root and ends at a terminal

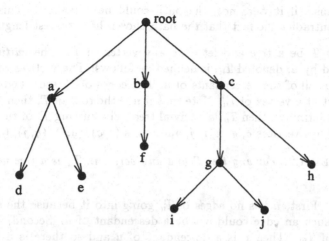

FIGURE 1.3. A tree.

vertex. If this were not true, we could extend the given path to a longer
one. The depth of a tree T is denoted $De(T)$. The depth of the tree in
Figure 1.3 is three because (root, c, g, j) is a path of greatest length.

Some general facts about trees are collected together in the following:

THEOREM 1.1. *Let T be a tree. Then we have:*

(1) *No vertex has more than one parent.*

(2) *If u and v are vertices of T and there is a path from u to v, then
there is no path from v to u.*

(3) *Every nonterminal vertex has a terminal descendant.*

PROOF.   (1) Suppose that w is a vertex with two parents u and
v. By definition of a tree, there is a path from the root to u.
Appending the edge (u, w) to this path produces a path from the
root to w. Similarly, appending the edge (v, w) to the path from
the root to v produces another path from the root to w. These
two paths are not the same because the last edge in one of them
is (u, w) and the last edge in the other is (v, w). This contradicts
the definition of a tree. Thus, w cannot have two parents.

(2) Suppose there is a path from v to u. Now start with the path from
the root to u and append to it first the path from u to v and then
the path from v back to u. This process produces a second path
from the root to u. This contradicts the definition of a tree. Thus,
there is no path from v to u.

(3) Let u be a nonterminal vertex. Consider a path of greatest length
starting at u. The vertex at the end of this path is terminal be-

cause, if it were not, the path could be extended. This would contradict the fact that the path chosen has greatest length. □

Now let T be a tree and let u be any vertex of T. The *cutting* of T determined by u, denoted T_u, is defined as follows: The vertices of T_u are u itself plus all of the descendants of u. The edges of T_u are all edges of T which start at a vertex of T_u. Note that if u is the root of T, then $T_u = T$, and if u is terminal, then T_u is a trivial tree. The cutting T_c of the tree in Figure 1.3 has vertices c, g, h, i, j, and edges (c, g), (g, i), (g, j), (c, h).

THEOREM 1.2. *For any tree T and any vertex u, T_u is a tree with u as the root.*

PROOF. First, u has no edges of T_u going into it because the starting vertex of such an edge could not be a descendant of u. Second, let v be a vertex of T_u. Then v is a descendant of u and so there is a path in T_u from u to v. The fact that there is only one such path follows from Exercise (14). □

Let T be a tree and u a vertex of T. The *quotient tree* T/u is defined as follows: The vertices of T/u are the vertices of T with the descendants of u removed; the edges of T/u are the edges of T which start and end at vertices of T/u. Thus, u is a terminal vertex of T/u. For example, in the tree of Figure 1.3, the quotient tree T/a is obtained by erasing the edges (a, d) and (a, e), and the vertices d and e. Notice that if u is the root of T, then T/u is trivial, and if u is terminal, then T/u is T. The proof of the following should be obvious and is omitted.

THEOREM 1.3. *If T is a tree and u is a vertex of T, then T/u is a tree whose root is the root of T.*

Finally, if T is a tree, then a *subtree* S of T is a tree whose vertices form a subset of the vertices of T; whose edges form a subset of the edges of T; whose root is the root of T; and whose terminal vertices form a subset of the terminal vertices of T. In the tree of Figure 1.3, the tree consisting of the root together with vertices c, h, g, j, and edges (root, c), (c, h), (c, g), (g, j) is a subtree.

The following theorem gives us an alternative definition of subtree. The proof is left as an exercise.

THEOREM 1.4. *Let S be a subtree of a tree T. Then S is the union of all the paths from the root to a terminal vertex of S. Conversely, if U is any nonempty subset of the set of all terminal vertices of the tree T, then the union of all paths from the root to a member of U is a subtree. The set of terminal vertices of this subtree is precisely U.*

For example, in the tree of Figure 1.3, consider the set of terminal vertices

$$U = \{d, f, j\}.$$

Then the subtree determined by U is the union of the three paths (root, a, d), (root, b, f), (root, c, g, j).

Exercises

(1) Sketch all cuttings of the tree in Figure 1.4.

(2) Sketch all quotient trees of the tree in Figure 1.4.

(3) How many subtrees does the tree in Figure 1.4 have?

(4) Sketch all cuttings of the tree in Figure 1.3.

(5) Sketch all quotient trees of the tree in Figure 1.3.

(6) How many subtrees does the tree in Figure 1.3 have?

(7) Prove that a tree has only one root.

(8) Define a *cycle* in a directed graph to be a path which begins and ends at the same vertex. Prove that trees do not contain cycles. Then prove that paths in trees consist of distinct vertices.

(9) Let G be a directed graph. For a vertex u of G, define $\rho(u)$ to be the number of edges starting at u minus the number of edges ending at u. Prove that

$$\sum_{u \in V(G)} \rho(u) = 0.$$

(10) Let D be a directed graph. Prove that the number of vertices u of D with $\rho(u)$ odd is even. [See Exercise (9) for the definition of $\rho(u)$.]

(11) Let T be a tree; let ϵ be the number of edges in T; and let ν be the number of vertices. Prove that $\epsilon = \nu - 1$.

(12) Let T be a nontrivial tree. Prove that there is a nonterminal vertex such that all of its children are terminal.

(13) Let T be a tree. Let W be a set of vertices of T such that every terminal vertex is in W and such that W contains a vertex whenever it contains all the children of that vertex. Prove that W contains all vertices.

(14) Let T be a tree and u and v vertices of T such that v is a descendant of u. Prove that there is only one path from u to v.

(15) Prove Theorem 1.4.

(16) Define a vertex w in a directed graph G to have the *unique path property* if, for any other vertex u in G, there is a unique path from w to u. Give an example of a directed graph with *two* vertices with the unique path property.

1.2. Game Trees

Let T be a nontrivial tree. We wish to use T to define an N-player game without chance moves. First, let $P_1, P_2, \ldots, P_N$ designate the players. Now

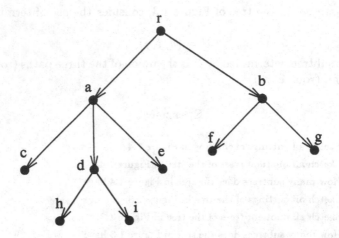

FIGURE 1.4. Tree for Exercises (1), (2),and (3) of Section 1.1.

label each nonterminal vertex with one of these designations. We will say that a vertex labeled with P_i *belongs* to P_i or that P_i *owns* that vertex. Then label each terminal vertex v with an N-tuple of numbers $\vec{p}(v)$. The game is now defined. It is played as follows. The player who owns the root chooses one of the children of the root. If that child is intermediate, the player to whom it belongs chooses one of *its* children. The game continues in this way until a terminal vertex v is reached. The players then receive payoffs in accordance with the N-tuple labeling this final vertex. That is, player P_i receives payoff $p_i(v)$, component number i of $\vec{p}(v)$. Figure 1.5 is an example with three players. In this game, there are never more than three edges out of a vertex. It is natural to designate these edges as L (for "left"), R (for "right"), and M (for "middle"). For example, if P_1's first move is to vertex b, then we would designate that edge (and move) by M. If P_3 then responds by moving to g, the *history* of that play of the game could be designated (M, R). As a second example, another possible history is (L, L, R). In this case, the payoff vector is $(0, -2, 2)$.

A tree labeled in the way just described (using designations of players and payoff vectors) is a *game tree*, and the corresponding game is a *tree game*.

It is important to realize that the kind of game we have set up has no chance moves. In a later section, we will see how to include the idea of vertices "belonging to chance" in our game trees. When that has been done, we will be able to think of all games as tree games.

Notice that every path from the root to a terminal vertex represents a different way in which the history of the game might evolve.

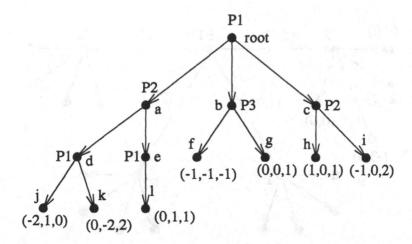

FIGURE 1.5. A three-player game tree.

It is not hard to see how a game tree might, in theory, be set up even for such a complicated game as chess. The player moving first would own the root. Each of her legal opening moves would correspond to a child of the root. Each of these children would belong to the player moving second and would have a child corresponding to each legal move for that player, etc. A terminal vertex would correspond to a situation in which the game is over such as a checkmate or a draw. Clearly, such a tree is enormous.

Let us consider a simpler example.

EXAMPLE 1.1 (TWO-FINGER MORRA). This game is played by two people as follows. Both players simultaneously hold up either one finger or two and, at the same time, predict the number held up by the other player by saying "one" or "two." If one player is correct in his prediction while the other player is wrong in hers, then the one who is right wins from the other an amount of money equal to the total number of fingers held up by both players. If neither is right or both are right, neither wins anything.

The difficulty in representing this game as a tree is the simultaneity of the players' moves. In a game defined by a tree the players obviously move consecutively. To remove this apparent difficulty, suppose that a neutral third party is present. Player P_1 whispers her move (that is, the number of fingers she wishes to hold up and her prediction) to the neutral person. Player P_2 then does the same. The neutral person then announces the result. Now the players' moves are consecutive.

We see that P_1 has a choice of four moves. Each can be represented by an ordered pair (f, p) where f is the number of fingers held up and p is

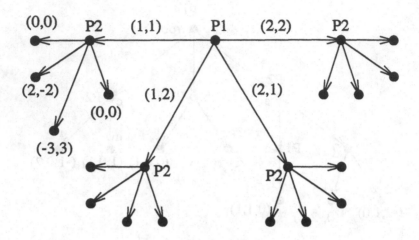

FIGURE 1.6. Two-finger morra.

the prediction. Thus each of f and p is either 1 or 2. We see from this that the root of our tree, which is labeled by P_1, has four children. These are all labeled by P_2 and each of them has four children (one for each of P_2's possible moves). These children are all terminal. The tree is shown in Figure 1.6. (Most of the payoff pairs have been left out.) This tree for two-finger morra does not embody all the rules of the game. It appears from the tree that P_2 could guarantee herself a positive payoff by making the correct move at any of her four vertices. This is certainly not true—the two players have symmetric roles in the game, and so if P_2 could guarantee herself a positive payoff, then so could P_1. But they cannot both win. The problem is that P_2 does not know P_1's move and thus does not know which of her vertices she is moving from. In other words, the game is not of perfect information. This concept was discussed briefly earlier, and will be discussed in more detail later.

Notice that if T is a game tree, then any cutting of T is also a game tree. A quotient tree T/u of T becomes a game tree if an N-tuple of numbers is assigned to the vertex u. For example, in Figure 1.5, the cutting T_a is a game in which P_3 has no vertices but receives payoffs. Also, T/a is a game if we assign, arbitrarily, the payoff $(0, 1, 1)$ to a.

A subtree of T is also a game tree. These constructions of new game trees from old ones are often useful in studying parts of games (such as end-games in chess) and restricted versions of games.

1.2.1. *Information Sets*

Let us go back and look at the game of Matching Coins (Figure 1.1). The rules of this game state that Louise does not know which coin Thelma is holding. This implies, of course, that the game is not of perfect information. In terms of the tree, it also means that Louise does not know which of her two vertices she is at. It is therefore impossible, under the rules, for her to plan her move (that is, decide which coin to hold) based on which of these two vertices she has reached. The set consisting of her two vertices is called an *information set*. More generally, an information set S for player P is a set of vertices, all belonging to P, such that, at a certain point in the game, P knows he is at one of the vertices in S but does not know which one.

We mentioned earlier that the tree in Figure 1.6 does not embody all the rules of two-finger morra. The reason is that the tree does not take account of the information set for P_2. This information set consists of all four of P_2's vertices.

Exercises

(1) For the game in Figure 1.5, sketch the cutting T_a. What is P_2's best move in T_a?

(2) For the game in Figure 1.5, sketch the quotient game T/a [with $(0,1,1)$ assigned to u]. What is P_1's best move?

(3) Concerning the game pictured in Figure 1.5, answer the following questions.
  • What is the amount player P_3 is guaranteed to win, assuming that all players play rationally?
  • What choice would you advise P_1 to make on his first move?
  • If the rules of the game allow P_1 to offer a bribe to another player, how much should he offer to whom for doing what?

(4) For the game tree in Figure 1.7, the three vertices owned by player B form an information set (enclosed by a dashed line). How should A and B play? How much can they expect to win?

(5) Suppose that, in the game of Figure 1.5, the rules are such that P_2 does not know whether P_1 has moved L or R. Sketch the tree with P_2's information set indicated. How should P_2 move?

(6) A very simple version of the game of nim is played as follows: There are two players, and, at the start, two piles on the table in front of them, each containing two matches. In turn, the players take any (positive) number of matches from *one* of the piles. The player taking the last match loses. Sketch a game tree. Show that the second player has a sure win.

(7) A slightly less simple version of nim is played as follows: There are two players, and, at the start, three piles on the table in front of them, each containing two matches. In turn, the players take any (positive) number of matches from *one* of the piles. The player taking the last match loses. Sketch a game tree. Show that the first player has a sure win.

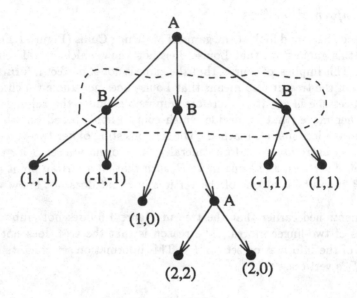

FIGURE 1.7. Game tree for Exercise (4) of Section 1.2.

1.3. Choice Functions and Strategies

We now want to make a precise definition of a "strategy" for a player in a game. *Webster's New Collegiate Dictionary* defines this word as "a careful plan or method." This is the general idea, but we need to make the idea rigorous.

There are three requirements which our definition of this concept must satisfy. The first is that it should be *complete*: A strategy should specify which move is to be made in every possible game situation. The second is that it should be *definite*: The move to be made in a given game situation should be determined by that game situation, and not by chance or by the whim of the player. It may help to think of a strategy as a description of how to play which can be implemented as a computer program. The third requirement is that *information sets must be respected*. This means that a strategy must call for the same move at every vertex in an information set. For example, in two-finger morra, a strategy for P_2 must call for the same move from each of her four vertices.

The following defines a concept which is our first approximation to what a strategy should be.

DEFINITION 1.2. Let T be a game tree and let P be one of the players. Define a *choice function for P* to be a function c, defined on the set of all vertices of T belonging to P, which is such that $c(u)$ is a child of u for every vertex u belonging to P.

Thus, given that player P is playing according to a choice function c, P knows which choice to make if the play of the game has reached vertex u (owned by P): She would choose $c(u)$.

Now suppose that there are N players $P_1, P_2, \ldots, P_N$. Let us denote the set of all choice functions for player P_i by Γ_i. Given that each player P_i moves according to his choice function $c_i \in \Gamma_i$, for $1 \leq i \leq N$, a path through the game tree from the root to a terminal vertex is determined. Define $\pi_i(c_1, c_2, \ldots, c_N)$ to be component number i of the N-tuple $\vec{p}(w)$ which labels the terminal vertex w at which this path ends. Thus, this quantity is the *payoff* to P_i when the players play according to the N-tuple of choice functions $(c_1, c_2, \ldots, c_N)$. For example, in the game of Figure 1.5, suppose that P_1, P_2, and P_3 use the choice functions c_1, c_2, and c_3, respectively, where $c_1(\text{root}) = a$, $c_1(d) = j$, $c_1(e) = l$; $c_2(a) = e$, $c_2(c) = i$; $c_3(b) = f$. The path traced out is (root, a, e, l), which terminates at the payoff vector $(0, 1, 1)$.

We can probably agree that for any strategy there is a choice function which embodies it. However, there are reasons why *choice function* is not an appropriate definition of *strategy*. The first is that choice functions are usually defined at many vertices where they need not be defined, that is, at vertices which can never be reached in the course of the game, given earlier decisions by the player. For example, in the game shown in Figure 1.5, a choice function for player P_1 which calls for him to move to the middle or right-hand child from the root obviously need not specify moves from either of the other vertices belonging to P_1 since they cannot be reached. Nevertheless the definition of a choice function requires that its domain of definition should consist of *all* vertices belonging to a given player.

1.3.1. *Choice Subtrees*

Let us try to eliminate the problem mentioned above. There are several ways in which this might be done. We could allow choice functions to be defined on *subsets* of the set of vertices belonging to a given player. This is unsatisfactory because the domain of definition would vary with the choice function. The following gives a better solution:

DEFINITION 1.3. Let T be a game tree and let P be one of the players. Let c be a choice function for P. Define a (P, c)-*path* to be a path from the root of T to a terminal vertex such that if u is a vertex of that path belonging to P, then the edge $(u, c(u))$ is an edge of the path.

Thus, a (P, c)-path represents one of the possible histories of the game given that player P plays according to the choice function c. For example, in the game of Figure 1.5, suppose that P_1 uses the choice function c_1, where $c_1(\text{root}) = a$, $c_1(d) = k$, $c_1(e) = l$. Then the path (root, a, e, l) is a (P_1, c_1)-path. So is (root, a, d, k). In fact, these are the only two.

Then we have the following:

DEFINITION 1.4. Let T be a game tree, P a player, and c a choice function for P. Then the *choice subtree determined by P and c* is defined to be the union of all the (P, c)-paths.

Thus, the choice subtree determined by P and c has as its set of vertices all the vertices of T which can be reached in the course of the game, given that P plays according to the choice function c. It is clear that a choice subtree is a subtree (by Theorem 1.4). The reader might now look at Figure 1.5 to see what the possible choice subtrees are. For example, one of them contains the vertices: root, a, e, d, k, and l. Its edges are (root, a), (a, d), (a, e), (d, k), and (e, l). It is interesting that if u is a vertex of a choice subtree which belongs to player P, then u has only one child in the subtree. On the other hand, a vertex v of a choice subtree which belongs to a player different from P has all its children in the subtree. (Both of these facts are proved in the next theorem.) We see that all the essential information about the choice function c is contained in the choice subtree. That is, if one knows the subtree one can play in accordance with the choice function. Suppose, for example, that player P is playing according to a choice subtree S. If the game has reached a vertex u belonging to P, then u is in S and only one of its children is in S. P's move is the edge from u to that child. On the other hand, the inessential information about the choice function (namely, how it is defined at vertices which can never be reached) is not contained in the choice subtree. For example, in the tree of Figure 1.5, consider two choice functions for P_1: $c_1(\text{root}) = b$, $c_1(d) = k$, $c_1(e) = l$; and $c_1'(\text{root}) = b$, $c_1'(d) = j$, $c_1'(e) = l$. These clearly embody the same strategy since vertex d is never reached. But the choice subtree determined by P_1 and c_1 is the same as the choice subtree determined by P_1 and c_1'.

Not all subtrees are choice subtrees. For example, in the game of Figure 1.5, the subtree whose terminal vertices are j, k, h, and i is not a choice subtree for any of the players. (If it were a choice subtree for P_1, it would not contain both j and k; if for P_2, it could not contain both h and i; and, if for P_3, it would contain either f or g.)

The two properties stated above for choice subtrees (concerning children of vertices) are enough to characterize them among all subtrees. This fact is contained in the following theorem. The advantage of this characterization is that we can recognize a subtree as a choice subtree without having to work with choice functions at all. A lemma is needed first.

LEMMA 1.5. *Let T be a game tree; let P be a player; and let c be a choice function for P. Suppose that Q is a path from the root to a vertex v and that Q satisfies the condition:*

(*) *If w is a vertex on Q owned by P, then either $w = v$ or $c(w)$ is on Q.*

Then Q can be extended to a (P, c)-path.

PROOF. Let Q' be a path of greatest length containing Q and satisfying (*). Let z be the last vertex on Q' (that is, the vertex furthest from the root). If z is not terminal, then either it belongs to P or it does not. If it does, let

$$Q'' = Q' \cup (z, c(z)).$$

If it does not, let y be any child of z and let

$$Q'' = Q' \cup (z, y).$$

Then Q'' satisfies (*) and is longer than Q'. This is a contradiction and so z is terminal. Hence, Q' is a (P, c)-path. $\square$

THEOREM 1.6. *Let T be a game tree and let P be one of its players. A subtree S of T is a choice subtree determined by P and some choice function c if and only if both the following hold:*

   (1) *If u is a vertex in S and u belongs to P, then exactly one of the children of u is in S.*
   (2) *If u is a vertex in S and u does not belong to P, then all the children of u are in S.*

PROOF. First, suppose S is a choice subtree determined by P and c. Let u be a vertex in S. If u belongs to P, then, by definition of choice subtree, there is a (P, c)-path containing u and contained in S. By definition of (P, c)-path, $(u, c(u))$ is an edge of this path and so $c(u)$ is in S. Now if v is a child of u different from $c(u)$, then no (P, c)-path contains v and thus v is not in S. Hence (1) holds.

Now if u does not belong to P, then let v be any child of u and let Q be the path from the root to v. The part of Q which ends at u is part of a (P, c)-path (since u is in S). Thus, Q satisfies (*) of Lemma 1.5 and so there is a (P, c)-path containing Q. Hence, v is in S.

For the other half of the proof, suppose that S is a subtree of T which satisfies both (1) and (2). We must define a choice function c for which S is the choice subtree determined by P and c. If v is a vertex of S belonging to P, define $c(v)$ to be that child of v which is in S. If v belongs to P but is not in S, let $c(v)$ be *any* child of v. Now we prove that this works. Let Q be a (P, c)-path. To show that Q is entirely contained in S, let u be the last vertex in Q which is in S. If u is terminal, then the entire path Q is in

S. Suppose then that u is not terminal. There are two cases: If u belongs to P, then $c(u)$ is in S and the edge $(u, c(u))$ is in Q. This contradicts the fact that u is the last vertex. This leaves only the case where u does not belong to P. But then if (u, v) is the edge from u which is in Q, then we again reach a contradiction since, by (2), v is in S.

We have now shown that S contains the choice subtree determined by P and c. To show the opposite inclusion, let w be a vertex of S. Then, since S is a subtree, the path from the root to w is contained in S. By (1) and the way in which c was constructed, this path satisfies condition (*) of Lemma 1.5. Hence there is a (P, c)-path containing w, and thus w is in S. $\square$

We now replace choice functions by choice subtrees. We will modify our earlier notation and write $\pi_i(S_1, \ldots, S_N)$ for the payoff to player P_i resulting from the N-tuple of choice subtrees $(S_1, \ldots, S_N)$.

Now, not every choice subtree respects information sets. Thus, we cannot simply define a strategy to be a choice subtree. Instead, we define a strategy for a player P to be a member of a subset of the set of all choice subtrees determined by P and P's choice functions. In turn, the members of this subset are the choice subtrees which respect P's information sets. We have the following:

DEFINITION 1.5. Let T be a game tree with N players

$$P_1, P_2, \ldots, P_N.$$

A *game in extensive form* based on T consists of T together with a non-empty set Σ_i of choice subtrees for each player P_i. The set Σ_i is called the *strategy set* for P_i and a member of Σ_i is called a *strategy* for P_i.

To summarize, it is this concept of strategy sets that allows us to impose the rules of the game which prohibit the players from having perfect information. For example, if P_i is not allowed to know which of a set of vertices he has reached at a certain point in the game, then P_i's strategy set would not contain choice subtrees which call for different moves depending on this unavailable information.

We will use an uppercase Greek letter Γ, Δ, or Λ to denote a game in extensive form. Formally, we would write, for example,

$$\Gamma = (T, \{P_1, P_2, \ldots, P_N\}, \{\Sigma_1, \Sigma_2, \ldots, \Sigma_N\})$$

to describe the game Γ with game tree T, and with players $P_1, P_2, \ldots, P_N$ having strategy sets $\Sigma_1, \Sigma_2, \ldots, \Sigma_N$, respectively.

Each choice subtree (and thus each strategy) is a subtree. Thus a choice subtree is itself a game tree. It represents a smaller game in which one of the players has effectively been removed (since that player has no choices).

Finally, notice that if a choice subtree S_i is chosen from each Σ_i, then the intersection of the S_i's is a path from the root to a terminal vertex. This path represents the history of the game, given that each player moves in accordance with the appropriate S_i. The N-tuple of numbers labeling the terminal vertex at which this path ends gives us the payoffs to the N players. We write $\pi_i(S_1, S_2, \ldots, S_N)$ for component number i of that N-tuple. This number is thus the payoff to player P_i resulting from all players playing according to the given choice subtrees.

Let us consider an example. Its game tree is given in Figure 1.8 and it is a game of perfect information. There are two players, designated A and B. Player A has twelve choice functions and player B has four. To count the choice functions for A, note that there are three vertices belonging to A. One of them (the root) has three children while each of the other two has two children. Thus, the number of choice functions is $2 \times 2 \times 3 = 12$. There are six choice subtrees for player A. Three of them are shown in Figure 1.9. To enumerate them, let M denote the choice subtree for A corresponding to A's choice of her middle child (at the root), and let R denote the one corresponding to the right-hand child. If A chooses her left-hand child at the root, there are four possible choice subtrees. Each of these has a name of the form Lxy, where x and y are each either L or R. In this notation, x denotes the child chosen by A in case B moves left, and y is the child chosen by A in response to B's moving right. Thus these four choice subtrees are LLL, LLR, LRL, and LRR.

For B, there are four choice subtrees. We denote them xy, where each of x and y is either L or R. Here, x is the response of B to A's choosing her left-hand child, and y is the response of B to A's choosing her middle child. One of these choice subtrees for B is shown in Figure 1.10. For an example, suppose that A plays according to LRR, while B plays according to LL. Then the payoff to A is 2, while the payoff to B is -1.

Now suppose that the rules of the game are modified so that, after A's first move, B cannot distinguish between A's moving left and A's moving middle. Also suppose that if A moves left on her first move, then A cannot distinguish between B's moving left and B's moving right. Then the strategy set for A consists of LLL, LRR, M, and R. Also, the strategy set for B consists of the two choice subtrees, LL and RR.

Exercises

(1) Describe all the strategies for each player for the version of nim given in Exercise (6) of Section 1.2.

(2) Describe all the strategies for each player in the version of nim given in Exercise (7) of Section 1.2.

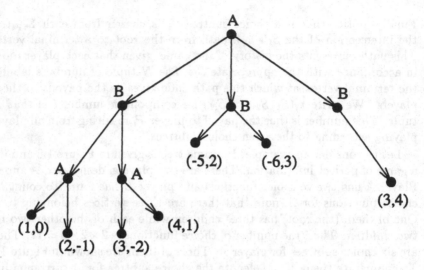

FIGURE 1.8. A game tree.

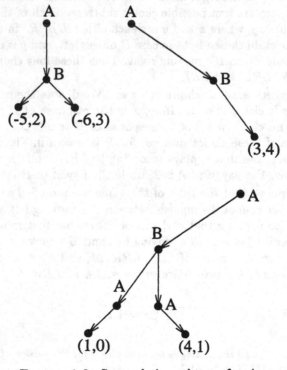

FIGURE 1.9. Some choice subtrees for *A*.

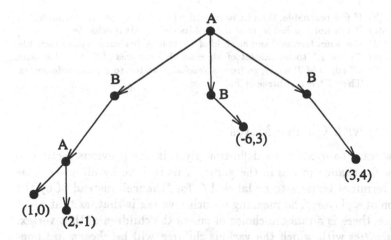

FIGURE 1.10. A choice subtree for B.

(3) Consider the game of perfect information shown in Figure 1.8. What strategies would you advise A and B to adopt?

(4) Suppose that, in the game of Figure 1.8, if A moves left at the root, she cannot distinguish between B's moving left and B's moving right. List the strategies for both players.

(5) In the game shown in Figure 1.8, suppose that the rules are such that, after A's first move, B cannot distinguish between A's moving left and A's moving middle. Also suppose that if A moves left on her first move, then A cannot distinguish between B's moving left and B's moving right. What strategies would you advise A and B to adopt?

(6) Sketch the choice subtrees for both players for the game in Figure 1.7. Which of these are strategies? (That is, which respect the information set?)

(7) In the game of Figure 1.5, how many choice subtrees are there for each player?

(8) Let T be a game tree with N players. Let P be one of the players and let c be a choice function for P. Define a vertex v of T to have the (P, c)-property if the path R from the root to v satisfies condition (*) of Lemma 1.5. Prove that the choice subtree determined by P and c is equal to the set of all vertices with the (P, c)-property.

(9) Let T be a game tree and let S be a choice subtree for player P. Let T_u be a cutting of T. Prove that either $S \cap T_u = \emptyset$ or $S \cap T_u$ is a choice subtree for P in T_u.

(10) Let
$$\Gamma = (T, \{P_1, \ldots, P_N\}, \{\Sigma_1, \ldots, \Sigma_N\})$$
be a game in extensive form. Define a vertex v of T to be *reachable* by a player P_i if there exists a strategy for P_i which contains v. A vertex v is simply *reachable* if it is reachable by every player. Prove the following statements.

   (a) The root is reachable.

(b) If v is reachable, then its parent (if v is not the root) is reachable.

(c) If v is not reachable, then none of its children is reachable.

(d) If v is not terminal and no child of v is reachable then v is not reachable.

(e) Define T^* to be the set of all reachable vertices of T together with all edges of T which go from a reachable vertex to a reachable vertex. Then T^* is a subtree of T.

1.4. Games with Chance Moves

We now want to modify the definition given in the previous section so as to allow for chance moves in the game. This is done by allowing some of the nonterminal vertices to be labeled C for "chance" instead of by the designation of a player. The meaning of such a vertex is that, at that point in the game, there is a random choice of one of the children of that vertex. The *probabilities* with which the various children will be chosen are non-negative numbers summing up to 1. These probabilities are used as labels on the edges coming out of the vertex labeled "chance." We introduce the following notation: If u is a vertex belonging to chance and v is a child of u, then $\Pr(u, v)$ denotes the probability that the edge (u, v) will be chosen. Thus, letting $E(u)$ denote the set of all edges starting at u,

$$\Pr(u, v) \geq 0$$

and

$$\sum_{(u,v) \in E(u)} \Pr(u, v) = 1.$$

For example, suppose that the game begins with a roll of two fair dice, and that the course of the rest of the game then depends on the sum of the numbers on top of the dice. Thus, there are eleven possible outcomes: $2, 3, \ldots, 12$, and so the root of the tree has eleven children. The probabilities on the edges coming out of the root vary from 1/36 (for outcomes 2 and 12) to 1/6 (for outcome 7).

The definition of choice function and of choice subtree is exactly the same as before. The definition of payoff $\pi_i(S_1, S_2, \ldots, S_N)$ for player P_i no longer works, however, because the terminal vertex where the game ends is now dependent on the chance moves as well as on the choices made by the players. Suppose that each player P_i plays according to choice subtree S_i. In the case of games without chance moves, the intersection of the S_i's is a single path from the root to a terminal vertex. In the case where there are chance moves, this intersection is a subtree which may contain many terminal vertices. The correct picture is that every path in this subtree branches whenever it runs into a vertex belonging to chance. Thus, every terminal vertex u in this intersection is reached with a probability which is the product of probabilities at the vertices belonging to chance encountered on the path from the root to u. The *expected* payoff to player P_i would then be a weighted average of ith components of the N-tuples labeling the terminal vertices of the intersection.

Our point of view here is that the game is played repeatedly, perhaps many thousands of times. It is clear that in a game which is only played once (or just a few times) the idea of expected payoff may have little meaning. We are making a rather important distinction since some games, by their nature, can only be played once. Consider, for example, games of nuclear confrontation in which payoffs may correspond to annihilation.

DEFINITION 1.6. Let T be a tree for a game with N players. For each i, let $S_i \in \Sigma_i$. If w is a terminal vertex of $\cap_{i=1}^N S_i$ and if R is the path from the root to w, then the probability of the game terminating at w is

$$\Pr(S_1, \ldots, S_N; w) = \prod \{\Pr(u, v) : u \text{ belongs to chance and } (u, v) \in R\}.$$

Here, it is understood that if there are no chance vertices on the path R, then $\Pr(S_1, \ldots, S_N; w) = 1$.

The payoff can then be defined as follows:

DEFINITION 1.7. In the setting of the previous definition, the *expected payoff*, $\pi_i(S_1, S_2, \ldots, S_N)$ to player P_i resulting from the choice subtrees $S_i \in \Sigma_i$ is defined by

$$\pi_i(S_1, \ldots, S_N) = \sum \Pr(S_1, \ldots, S_N; w) p_i(w),$$

where the sum is taken over all terminal vertices w in $\cap_{j=1}^N S_i$.

Let us do some calculating for the game shown in Figure 1.11. There are two players, designated A and B, and the game is of perfect information. There are two vertices belonging to chance. The left-hand vertex has two children, and the probabilities attached to the two edges are 0.6 and 0.4, respectively. One might imagine that a biased coin is tossed at that point in the game. The right-hand chance vertex is similar except that the coin tossed is fair. Player A has only two strategies: He can either move left (denoted L) or right (denoted R). The reader is invited to verify that player B has sixteen strategies.

Now suppose that player A plays his strategy L, while player B plays her strategy $LRLR$. The meaning of this notation is that if A moves left and the chance move is then to the left child, B moves left. If A moves left and the chance move is then to the right, B moves right. Finally, if A moves right, then B moves left or right depending on whether the chance move is to the left or right. To compute the expected payoffs for both players, note first that the terminal vertices which can be reached, given that the

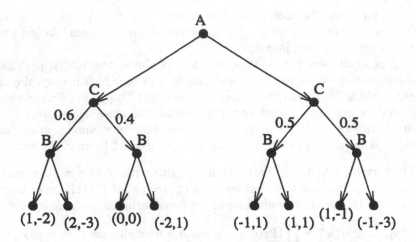

FIGURE 1.11. A game with chance moves.

players play according to these strategies, are the ones whose payoff pairs are $(1, -2)$ and $(-2, 1)$. Thus, the expected payoff for player A is

$$\pi_A(L, LRLR) = 0.6 \times 1 + 0.4 \times -2 = -0.2,$$

while the expected payoff for player B is

$$\pi_B(L, LRLR) = 0.6 \times (-2) + 0.4 \times 1 = -0.8.$$

1.4.1. *A Theorem on Payoffs*

We now wish to prove a theorem which allows us to compute payoffs due to choice subtrees in terms of payoffs in smaller games. A lemma is needed first.

LEMMA 1.7. *Let T be a tree game with N players; let $S_1, \ldots, S_N$ be choice subtrees for $P_1, \ldots, P_N$, respectively; and let r be the root of T. Then we have:*

(1) *If r belongs to a player P_i and (r, u) is an edge of S_i, then, for $1 \le j \le N$, $S_j \cap T_u$ is a choice subtree of T_u for player P_j.*

(2) *If r belongs to chance and u is any child of r, then, for $1 \le j \le N$, $S_j \cap T_u$ is a choice subtree of T_u for player P_j.*

PROOF. By Exercise (9) of Section 1.3, it suffices for the proof of both (1) and (2) to show that $S_j \cap T_u \neq \emptyset$. This is true in both cases because $u \in S_j$ for $1 \le j \le N$. $\square$

THEOREM 1.8. *Let T be a tree game with N players $P_1, \ldots, P_N$. Let $S_1, \ldots, S_N$ be choice subtrees for $P_1, \ldots, P_N$, respectively. Then, with r as the root of T, we have:*

(1) *If r belongs to the player P_i and (r, u) is in S_i then, for $1 \leq j \leq N$,*

$$\pi_j(S_1, S_2, \ldots, S_N) = \pi_j(S_1 \cap T_u, S_2 \cap T_u, \ldots, S_N \cap T_u).$$

(2) *If r belongs to chance then, for $1 \leq j \leq N$,*

$$\pi_j(S_1, S_2, \ldots, S_N) = \sum_{u \in Ch(r)} \Pr(r, u) \pi_j(S_1 \cap T_u, \ldots, S_N \cap T_u).$$

PROOF. To prove (1), note that each $S_k \cap T_u$ is a choice subtree of T_u. This follows from Lemma 1.7. Then, notice that $\cap_{k=1}^{N} S_k$ is $\cap_{k=1}^{N}(S_k \cap T_u)$ together with the edge (r, u). Thus, the subtree $\cap_{k=1}^{N} S_k$ of T and the subtree $\cap_{k=1}^{N}(S_k \cap T_u)$ of T_u have the same terminal vertices. Also the probability of the game terminating at each of these terminal vertices is the same in both game trees. The result follows from the definition of expected payoff.

To prove (2), note that Lemma 1.7 implies that $S_k \cap T_u$ is a choice subtree of T_u for each k and each child u of r. Note also that

$$\Pr(S_1, \ldots, S_N; w) = \Pr(r, u) \Pr(S_1 \cap T_u, \ldots, S_N \cap T_u; w),$$

for any terminal vertex $w \in \cap_{k=1}^{N}(S_k \cap T_u)$. Then, writing π_j for

$$\pi_j(S_1, \ldots, S_N)$$

and summing over all terminal vertices $w \in \cap_{k=1}^{N} S_k$, we have

$$
\begin{aligned}
\pi_j &= \sum \Pr(S_1, \ldots, S_N; w) p_j(w) \\
&= \sum_{u \in Ch(r)} \sum_{w \in \cap(S_k \cap T_u)} \Pr(S_1, \ldots, S_N; w) p_j(w) \\
&= \sum_{u \in Ch(r)} \Pr(r, u) \sum_{w \in \cap(S_k \cap T_u)} \Pr(S_1 \cap T_u, \ldots, S_N \cap T_u; w) p_j(w) \\
&= \sum_{u \in Ch(r)} \Pr(r, u) \pi_j(S_1 \cap T_u, \ldots, S_N \cap T_u). \quad \square
\end{aligned}
$$

Exercises

(1) For the game of perfect information presented in Figure 1.11, what is your advice to player A about how to play? What about player B?

(2) The game tree shown in Figure 1.12 has one chance move; it is not of perfect information (the information sets are enclosed by dashed lines.) List the strategies for each of the two players.

(3) Suppose that the rules for the game shown in Figure 1.11 are changed so that B does not know the result of either one of the chance moves. What is your advice to player A about how to play? What about player B?

(4) Two players, call them Frankie and Johnny, play the following card game (which we will call One Card): First, each antes $1 (that is, puts it into the "pot" in the middle of the table). Then a card is dealt face-down to Frankie from a deck which contains only cards marked *high* and *low* (in equal numbers). Frankie looks at her card and bets either $1 or $5 (by putting the money into the pot). Whichever the bet, Johnny can either *see* (by matching the bet) or *fold*. If Johnny folds, Frankie takes all the money in the pot. If he sees, the card is revealed. Then, if it is high, Frankie takes all the money and, if low, Johnny takes it all. Write down a game tree for One Card, indicating the information sets.

(5) The game of Sevens is played between two players (called A and B) as follows. Each player rolls a fair die in such a way that the other cannot see which number came up. Player A must then bet $1 that either: (i) The *total* on the two dice is less than seven, or (ii) the total is greater than seven. Then, B can either (i) accept the bet, or (ii) reject it. If the bet is rejected, the payoff to each is zero. Otherwise, both dice are revealed. If the total is exactly seven, then both payoffs are zero. Otherwise, one of the players wins the other's dollar. Describe a tree for Sevens, including information sets.

(6) Let T be an N-player game tree. Given a choice subtree S_i for each i, let $H = \cap_{i=1}^{N} S_i$ and prove that either
   (a) H is a single path from the root to a terminal vertex; or
   (b) every path in H from the root to a terminal vertex contains a vertex belonging to chance.

1.5. Equilibrium N-tuples of Strategies

Roughly speaking, an N-tuple of strategies (one for each player) is in equilibrium if one player's departing from it while the others remain faithful to it results in punishment for the straying player. The idea is that once the players start playing according to such an N-tuple, then they all have good reason to stay with it. This gives us a "solution concept," one which is reasonably amenable to mathematical analysis. There is a large class of games (to be considered in the next chapter) for which everyone agrees that an equilibrium N-tuple really is a "solution." For other games, this is not so clear. We will discuss this question later. The formal definition is the following:

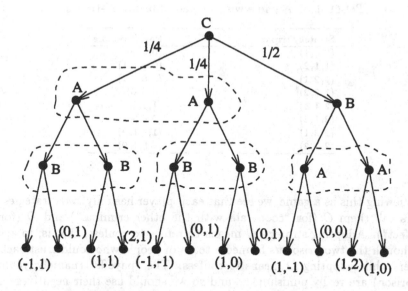

FIGURE 1.12. Game tree for Exercise (2) of Section 1.4.

DEFINITION 1.8. Let Γ be an N-player game in extensive form and denote the players' strategy sets by $\Sigma_1, \Sigma_2, \ldots, \Sigma_N$. An N-tuple (S_i^*), where

$$S_i^* \in \Sigma_i$$

for $1 \leq i \leq n$, is an *equilibrium N-tuple* (or is *in equilibrium*) if, for every i and for every $S_i \in \Sigma_i$, we have

$$\pi_i(S_1^*, \ldots, S_i, \ldots, S_N^*) \leq \pi_i(S_1^*, \ldots, S_i^*, \ldots, S_N^*).$$

Thus, a single player who departs from the N-tuple is hurt (or, at least, not helped). It is easy to check that the pair of strategies $(L, LRLR)$ in the game shown in Figure 1.11 is in equilibrium. The following game is an important example in this context.

EXAMPLE 1.2 (THE PRISONER'S DILEMMA). Two criminals, call them Bonnie and Clyde, are arrested for robbery. The police immediately separate them so that they are unable to communicate in any way. Each is offered the following deal: If you confess and implicate the other prisoner, then you will serve only one year in jail (if the other guy doesn't confess) or five years (if the other guy does confess). On the other hand, if you don't confess and the other prisoner does, you will serve ten years. Finally, if neither of you confesses, you will both serve two years since our case won't be very strong.

TABLE 1.1. A game with two equilibrium N-tuples.

Strategy triples	Payoff vectors
$(1,1,1)$	$(0,0,0)$
$(1,1,2)$	$(0,-1,1)$
$(1,2,1)$	$(1,0,-1)$
$(1,2,2)^\sharp$	$(0,0,0)$
$(2,1,1)^\sharp$	$(1,1,-2)$
$(2,1,2)$	$(2,0,-2)$
$(2,2,1)$	$(1,-1,0)$
$(2,2,2)$	$(-1,1,0)$

Viewing this as a game, we see that each player has only two strategies. Let's call them C (for "cooperate with the other criminal") and D (for "defect"). Adopting strategy C means to refuse to confess, that is, to act as though the two prisoners formed a team of people who could trust each other, while adopting D means to confess. The "payoffs" (namely, years in prison) are really punishments and so we should use their negatives as payoffs instead. We see that the payoff to each player arising from the pair of strategies (D, D) is -5. For the strategy pair (D, C) (that is, Bonnie defects while Clyde cooperates), Bonnie "wins" -1 and Clyde "wins" -10. Now, it is easy to see that (D, D) is an equilibrium pair. Moreover, it is the only one. Nevertheless, it is a little difficult to think of (D, D) as the "solution" to Prisoner's Dilemma. They both do better if they play the strategy pair (C, C). Of course, they would have to trust each other in order to do so. This game will be discussed again in a later chapter.

Another example is described in Table 1.1. In this game, there are three players. Each has only two strategies, denoted 1 and 2. The players simultaneously choose their strategies, so that no player knows what either of the other two chose. Thus, the game is not of perfect information. There are eight combinations of strategies. These combinations are listed in the left-hand column, and the corresponding 3-tuples of payoffs are in the right-hand column. The equilibrium 3-tuples are marked with $\sharp$.

We see that there are two equilibrium 3-tuples, neither of which looks very stable. It is clear that the third player would strongly prefer the 3-tuple $(1, 2, 2)$, while the other two players would prefer the 3-tuple $(2, 1, 1)$. These two could force the issue by playing according to 2 and 1, respectively. However, it would be worth his while for player 3 to make a payment to player 2 in exchange for 2's playing strategy 2.

Despite our reservations about whether knowing an equilibrium N-tuple solves a game, the concept is still of value. It is therefore of interest to know conditions under which one exists. It is certainly true that there are many games for which there is no equilibrium N-tuple. The reader may verify that this is true for two-finger morra. The most important class of games for which existence is guaranteed is that of games of perfect information. This type of game was defined earlier, and a definition can now be given in terms of our notation for games in extensive form.

DEFINITION 1.9. Let Γ be a game in extensive form. Then Γ is *of perfect information* if, for every player, each choice subtree is a strategy.

We can now state the following:

THEOREM 1.9. *Let Γ be a game in extensive form. If Γ is of perfect information, then there exists an equilibrium N-tuple of strategies.*

PROOF. The proof is by induction on the depth of the game tree T. The smallest possible depth of a game tree is one, since we excluded trivial trees. Now, if $\text{De}(T) = 1$, then at most one player has a move (depending on whether the root belongs to a player or to chance). Assuming that the root belongs to player P, let the strategy for P be to choose a (terminal) child of the root for which that player's payoff is a maximum. The corresponding choice subtree for P then consists of a single edge. Let the choice subtrees for the other players be all of T (in fact, no other possibilities exist). That the resulting N-tuple is in equilibrium should be clear. In the case where the root belongs to chance, each player's choice subtree must be taken to be all of T (since no other possibilities exist). The proof is complete in case $\text{De}(T) = 1$.

Now suppose that the depth of T is $m > 1$ and that the theorem holds for all games of perfect information for which the depth of the game tree is less than m. Let r be the root of T. For each child u of r, the cutting T_u is a game tree and $\text{De}(T_u) < \text{De}(T)$. We regard T_u as the game tree for a game of perfect information by defining each player's strategy set to be the set of all choice subtrees determined by that player. By the inductive assumption, there is an equilibrium N-tuple, $(S_1^u, S_2^u, \ldots, S_N^u)$, of strategies in T_u. We want to put these together to form an N-tuple of strategies in T. There are two cases. First, if the root of T belongs to a player P_j, choose u to be a child of the root such that the payoff $\pi_j(S_1^u, S_2^u, \ldots, S_N^u)$ is a maximum over $\text{Ch}(r)$. Now define S_j^* to be the union of S_j^u and the edge (r, u). For $1 \leq i \leq N$ and $i \neq j$, define S_i^* to be

$$\bigcup_{v \in \text{Ch}(r)} ((r, v) \cup S_i^v).$$

Second, if r belongs to chance, define each S_i^* according to the preceding formula.

It should be clear that we have now defined an N-tuple of choice subtrees. Since the game is of perfect information, all these choice subtrees are strategies. We must now show that this N-tuple of strategies is in equilibrium. To do so, suppose that one of the players, say P_i, plays according to strategy S_i instead of S_i^*. We consider the following three cases:

(1) Suppose the root belongs to player P_i. Now if (r, w) is in S_i, we have, by Theorem 1.8,

$$\pi_i(S_1^*, \ldots, S_i, \ldots, S_N^*) = \pi_i(S_1^* \cap T_w, \ldots, S_i \cap T_w, \ldots, S_N^* \cap T_w). \quad (1.1)$$

Then since $(S_1^w, \ldots, S_N^w)$ is an equilibrium N-tuple in T_w, we have

$$\pi_i(S_1^* \cap T_w, \ldots, S_i \cap T_w, \ldots, S_N^* \cap T_w) \leq \pi_i(S_1^w, \ldots, S_N^w).$$

Now if the edge (r, u) belongs to S_i^*, we have, by the way in which S_i^* was chosen,

$$\pi_i(S_1^w, \ldots, S_N^w) \leq \pi_i(S_1^u, \ldots, S_N^u).$$

By definition of the S_i^*'s, we have

$$\pi_i(S_1^* \cap T_u, \ldots, S_N^* \cap T_u) = \pi_i(S_1^u, \ldots, S_N^u).$$

Finally, by Theorem 1.8,

$$\pi_i(S_1^*, \ldots, S_N^*) = \pi_i(S_1^* \cap T_u, \ldots, S_N^* \cap T_u).$$

Combining all these, we get

$$\pi_i(S_1^*, \ldots, S_i, \ldots, S_N^*) \leq \pi_i(S_1^*, \ldots, S_N^*).$$

(2) Suppose that the root belongs to a player P_j different from P_i. Let u be that child of the root so that (r, u) is in S_j^*. Then we have

$$\pi_i(S_1^*, \ldots, S_i, \ldots, S_N^*) = \pi_i(S_1^* \cap T_u, \ldots, S_i \cap T_u, \ldots, S_N^* \cap T_u).$$

But then

$$\pi_i(S_1^* \cap T_u, \ldots, S_i \cap T_u, \ldots, S_N^* \cap T_u) \leq \pi_i(S_1^u, \ldots, S_N^u).$$

Finally,

$$\pi_i(S_1^u, \ldots, S_N^u) = \pi_i(S_1^*, \ldots, S_N^*).$$

Combining these yields the desired inequality.

TABLE 1.2. Game for Exercise (4) of Section 1.5.

Strategy triples	Payoff vectors
(1,1,1)	(1,-1,1)
(1,1,2)	(0,0,0)
(1,2,1)	(-1,2,0)
(1,2,2)	(0,1,-1)
(2,1,1)	(1,1,-2)
(2,1,2)	(-2,1,0)
(2,2,1)	(1,0,1)
(2,2,2)	(0,0,1)

(3) Suppose that the root belongs to chance. Then, for each child u of the root, we have

$$\pi_i(S_1^* \cap T_u, \ldots, S_i \cap T_u, \ldots, S_N^* \cap T_u) \leq \pi_i(S_1^* \cap T_u, \ldots, S_N^* \cap T_u).$$

Now apply part (2) of Theorem 1.8 to get that

$$\pi_i(S_1^*, \ldots, S_i, \ldots, S_N^*) \leq \pi_i(S_1^*, \ldots, S_N^*).$$

$\square$

This theorem was essentially proved for the first time in [vNM44].

Exercises

(1) Find all equilibrium pairs of strategies (if any) of the game in Figure 1.7.

(2) Determine the equilibrium pairs of strategies for the game of One Card (see Exercise (4) of Section 1.4).

(3) Verify that there are no equilibrium pairs of strategies for two-finger morra.

(4) Table 1.2 describes a three-person game in the way that Table 1.1 does. Are there any equlibrium 3-tuples for this game? Sketch a tree for this game, indicating information sets.

1.6. Normal Forms

Let

$$\Gamma = (T, \{P_1, \ldots, P_N\}, \{\Sigma_1, \ldots, \Sigma_N\})$$

be a game in extensive form. For every N-tuple of strategies $(S_1, \ldots, S_N)$ in the Cartesian product $\Sigma_1 \times \cdots \times \Sigma_N$, there is determined an N-tuple $\vec{\pi}(S_1, \ldots, S_N)$ of payoffs to the players. This function from $\Sigma_1 \times \cdots \times \Sigma_N$ to $\Re^N$ (that is, the N-fold Cartesian product of the real line $\Re$ with itself) is called the *normal form* of the game Γ. We will often study games in their normal forms without ever mentioning their extensive forms.

It should be mentioned that the normal form of a game is not uniquely defined. If we permute the order of the players then we change the Cartesian product $\Sigma_1 \times \cdots \times \Sigma_N$ and thus the function $\vec{\pi}$. In practice, this lack of uniqueness presents no difficulty and we will ignore it. Thus when we say "the normal form" we will mean any of the possible ones.

We abstract the idea of the normal form of a game in the following:

DEFINITION 1.10. Let $X_1, \ldots, X_N$ be finite nonempty sets and let $\vec{\pi}$ be a function from the Cartesian product $X_1 \times \cdots \times X_N$ to $\Re^N$. Then $\vec{\pi}$ is called an N-player game in normal form with strategy sets $X_1, \ldots, X_N$.

Let us designate the players in an N-player game $\vec{\pi}$ in normal form by $P_1, \ldots, P_N$. They play as follows: Each player P_i chooses $x_i \in X_i$. These choices are made simultaneously and independently. The payoffs are then given by the components of the N-tuple $\vec{\pi}(\vec{x})$, where

$$\vec{x} = (x_1, \ldots, x_N).$$

The definition of equilibrium N-tuples carries over almost unchanged to the context of games in normal form. To be precise, we give

DEFINITION 1.11. Let $\vec{\pi}$ be a game in normal form with strategy sets $X_1, \ldots, X_N$. An N-tuple

$$\vec{x}^* \in X_1 \times \cdots \times X_N$$

is an *equilibrium N-tuple* if, for all $1 \leq i \leq N$ and any $x_i \in X_i$,

$$\pi_i(x_1^*, \ldots, x_i, \ldots, x_N^*) \leq \pi_i(\vec{x}^*).$$

Certainly it is true that if a game in extensive form has an equilibrium N-tuple of strategies, then so does its normal form.

In the case where $N = 2$, a game in normal form can be pictured as a pair of matrices. In fact, let m be the number of strategies in Σ_1 and let n be the number of strategies in Σ_2. Form an $m \times n$ matrix of payoffs for player P_1 by labeling the rows with members of Σ_1 and the columns with members of Σ_2. The entry in the row labeled with $x \in \Sigma_1$ and in the column labeled with $y \in \Sigma_2$ is defined to be $\pi_1(x, y)$. Call this matrix M^1. The payoff matrix M^2 for player P_2 is formed in a similar way.

These matrices are not unique since a permutation of either or both of the sets Σ_1 or Σ_2 would also permute the rows or columns of the matrices. This lack of uniqueness presents no difficulty.

Let us compute the normal form for the two-person game of perfect information whose tree is shown in Figure 1.13. The players are designated A and B. There are eight strategies for player A:

$$LLL, LLR, LRL, LRR, RLL, RLR, RRL, RRR.$$

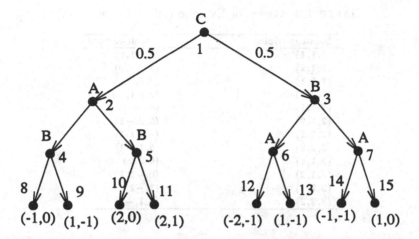

FIGURE 1.13. A two-person game.

Here again, L means "left" and R means "right." Thus, the strategy RLR corresponds to the choice function c_A, where

$$c_A(2) = 5, \quad c_A(6) = 12, \quad c_A(7) = 15.$$

Player B also has eight strategies and they are designated in exactly the same way. For example, B's strategy LLR corresponds to the choice function c_B, where

$$c_B(3) = 6, \quad c_B(4) = 8, \quad c_B(5) = 11.$$

The normal form of this game is contained in the following matrices. The order of the rows and columns corresponds to the order of strategies given above. The matrix M^1 of payoffs for player P_1 is given first.

$$\begin{pmatrix}
-3/2 & -3/2 & -1/2 & -1/2 & -1 & -1 & 0 & 0 \\
-3/2 & -3/2 & -1/2 & -1/2 & 0 & 0 & 1 & 1 \\
0 & 0 & 1 & 1 & -1 & -1 & 0 & 0 \\
0 & 0 & 1 & 1 & 0 & 0 & 1 & 1 \\
0 & 0 & 0 & 0 & 1/2 & 1/2 & 1/2 & 1/2 \\
0 & 0 & 0 & 0 & 3/2 & 3/2 & 3/2 & 3/2 \\
3/2 & 3/2 & 3/2 & 3/2 & 1/2 & 1/2 & 1/2 & 1/2 \\
3/2 & 3/2 & 3/2 & 3/2 & 3/2 & 3/2 & 3/2 & 3/2
\end{pmatrix}.$$

Then the payoff matrix M^2 for player P_2 is

TABLE 1.3. Game for Exercise (10) of Section 1.6.

Strategy triples	Payoff vectors
(1,1,1)	(0,−1,0)
(1,1,2)	(0,−2,0)
(1,2,1)	(3,0,−1)
(1,2,2)	(1,−1,−1)
(2,1,1)	(0,0,0)
(2,1,2)	(0,0,−1)
(2,2,1)	(−1,1,1)
(2,2,2)	(2,1,−1)
(3,1,1)	(0,0,2)
(3,1,2)	(0,−1,1)
(3,2,1)	(1,−2,1)
(3,2,2)	(1,1,−1)

$$
\begin{pmatrix}
-1/2 & -1/2 & -1 & -1 & -1/2 & -1/2 & -1 & -1 \\
-1/2 & -1/2 & -1 & -1 & 0 & 0 & -1/2 & -1/2 \\
-1/2 & -1/2 & -1 & -1 & -1/2 & -1/2 & -1 & -1 \\
-1/2 & -1/2 & -1 & -1 & 0 & 0 & -1/2 & -1/2 \\
-1/2 & 0 & -1/2 & 0 & -1/2 & 0 & -1/2 & 0 \\
-1/2 & 0 & -1/2 & 0 & 0 & 1/2 & 0 & 1/2 \\
-1/2 & 0 & -1/2 & 0 & -1/2 & 0 & -1/2 & 0 \\
-1/2 & 0 & -1/2 & 0 & 0 & 1/2 & 0 & 1/2
\end{pmatrix}.
$$

From the two matrices, it is easy to pick out the equilibrium pairs of strategies. Indeed, it is clear that a pair (S_1^*, S_2^*) is an equilibrium pair if and only if the entry at coordinates (S_1^*, S_2^*) in M^1 is a maximum in its column, while the entry in M^2 at the same coordinates is a maximum in its row.

It can be observed that there are six equilibrium pairs in this example.

Both the extensive and normal forms of games have advantages. The normal form is perhaps simpler mathematically. For the kind of games to be considered in the next chapter, the normal form is the one which allows us to solve the games. On the other hand, it is simpler to go from a verbal description of a game to its extensive form than to its normal form. Also, if we are interested in considering small subgames of a game as a way of better understanding a game which is too large to analyze, then the extensive form is better. This is because it is fairly easy to determine which small pieces of the big tree to examine. It is difficult to break down a normal form into pieces in this way. This sort of analysis will come up in Chapter 7.

Exercises

(1) Choose five entries in each of the two 8 × 8 matrices presented in this section and verify them.

(2) Give the normal form (as a pair of payoff matrices) of the game shown in Figure 1.7.

(3) Write down the normal form for the game shown in Figure 1.12. Find the equilibrium pairs (if any).

(4) Find all the equilibrium pairs of strategies for the game discussed in this section.

(5) Write down the normal form for the game shown in Figure 1.5.

(6) Write down the normal form for the game shown in Figure 1.8.

(7) Write down the normal form for the game shown in Figure 1.11.

(8) The game of Fingers is played as follows: The two players (Mary and Ned) simultaneously hold up either one or two fingers. Mary wins in case of a *match* (the same number), and Ned wins in case of a nonmatch. The amount won is the number of fingers held up by the winner. It is paid to the winner by the loser.
   (a) Describe the strategies for both players.
   (b) Write down the normal form (as a pair of payoff matrices).
   (c) Verify that there is no equilibrium pair of strategies.

(9) Write down the normal form of two-finger morra (see Figure 1.6).

(10) The normal form of a three-player game is given in Table 1.3. Player P_1 has three strategies (1, 2, 3); players P_2 and P_3 each have two (1 and 2). Find the two equilibrium 3-tuples of strategies.

(11) Refer to Exercise (10) of Section 1.3 for the terminology and notation of this exercise. Define
$$\Gamma^* = (T^*, \{P_1, \dots, P_N\}, \{\Sigma_i^*, \dots, \Sigma_N^*\}),$$
where $\Sigma_i^* = \{S_i \cap T^* : S_i \in \Sigma_i\}$. Prove that Γ and Γ^* have the same normal form.

2
Two-Person Zero-Sum Games

Let $\bar{\pi}$ be a game in normal form with strategy sets $X_1, \ldots, X_N$. We say that this game is *zero-sum* if

$$\sum_{i=1}^{N} \pi_i(x_1, \ldots, x_N) = 0,$$

for every choice of $x_i \in X_i, 1 \le i \le N$. The corresponding definition for a game in extensive form states that the sum of the components of $\vec{p}(v)$ is zero for each terminal vertex v. This condition is certainly true for ordinary recreational games. It says that one player cannot win an amount unless the other players jointly lose the same amount. Nonrecreational games, however, tend not to be zero-sum. Competitive situations in economics and international politics are often of the type where the players can jointly do better by playing appropriately, and jointly do worse by playing stupidly. The phrase "zero-sum game" has entered the language of politics and business.

In this chapter we are concerned with games in normal form which are zero-sum and which have two players. In the previous chapter, we discussed the fact that a two-person game in normal form can be represented as a pair of payoff matrices (one for each player). If the game is also zero-sum, the two matrices are clearly just negatives of each other. For this reason, there is no need to write down both. From now on, we shall represent a two-person zero-sum game as a *matrix game*, that is, as a single $m \times n$

matrix M. The two players are referred to as the *row player* and the *column player*, respectively. The row player has m strategies which are identified with the rows of M. The column player has n strategies which are identified with the columns of M. If the row player plays strategy i and the column player plays strategy j, then the payoff to the row player is m_{ij} and the payoff to the column player is $-m_{ij}$.

It is important to make it clear from the beginning that larger numbers in M are favored by the row player and smaller ones by the column player. Thus, a negative entry is a loss to the row player but a gain (of the absolute value) to the column player.

2.1. Saddle Points

The idea of an equilibrium pair of strategies can be easily translated into the context of matrix games. Indeed, if (p, q) is such a pair, then

$$m_{iq} \leq m_{pq} \text{ for all } i,$$

and

$$m_{pj} \geq m_{pq} \text{ for all } j.$$

Notice that the second inequality is reversed because the payoff to the column player is the negative of the matrix entry. These inequalities motivate the following definition.

DEFINITION 2.1. Let M be a matrix with real entries. An entry m_{pq} of M is a *saddle point* of M if it is simultaneously a minimum in its row and a maximum in its column.

Thus, if M is a matrix game, then m_{pq} is a saddle point if and only if (p, q) is an equilibrium pair of strategies. In the following three examples, the entry m_{21} is the only saddle point of the first matrix, entry m_{12} and entry m_{33} are both saddle points of the second matrix (there are two more), and the third matrix has no saddle points at all.

$$\begin{pmatrix} -2 & 3 \\ -1 & 1 \end{pmatrix} \qquad \begin{pmatrix} 2 & 1 & 1 \\ -1 & 0 & -1 \\ 3 & 1 & 1 \end{pmatrix} \qquad \begin{pmatrix} -1 & 0 & 1 \\ 1 & 2 & 3 \\ 2 & -1 & 1 \end{pmatrix}$$

The first thing we prove about saddle points is the following:

THEOREM 2.1. *If m_{kl} and m_{pq} are saddle points of the matrix M, then m_{kq} and m_{pl} are also saddle points and*

$$m_{kl} = m_{pq} = m_{kq} = m_{pl}.$$

PROOF. Since saddle points are maxima in their columns and minima in their rows, we have

$$m_{kl} \leq m_{kq} \leq m_{pq},$$

while

$$m_{pq} \leq m_{pl} \leq m_{kl}.$$

Thus,

$$m_{kl} = m_{pq} = m_{kq} = m_{pl}.$$

Also, m_{kq} is a saddle point since $m_{kq} = m_{kl}$ is a minimum in row k, and $m_{kq} = m_{pq}$ is a maximum in column q. Similarly, m_{pl} is a saddle point. □

We want to establish that a saddle point provides an acceptable solution to a matrix game. To do so, let us consider how the row player should rationally play. First, it is reasonable to assume that, no matter how he plays, his opponent will respond by playing so as to maximize her (the column player's) payoff. After all, she knows the game as well as he does and it would be foolish to assume that she will, in the long run, go on being charitable or making mistakes. Now, the fact that the game is zero-sum means that maximization of the column player's payoff is precisely the same as minimizing the row player's payoff. Therefore, the row player should choose his strategy so that the minimum possible payoff due to this strategy is as large as possible. The column player should act in a similar way. These ideas lead to the following:

DEFINITION 2.2. Let M be an $m \times n$ matrix game. The *value to the row player* and *value to the column player* are, respectively,

$$u_r(M) = \max_i \min_j m_{ij}$$

and

$$u_c(M) = \min_j \max_i m_{ij}.$$

Thus, $u_r(M)$ is an amount that the row player is *guaranteed* to win if he plays a strategy k for which the maximum in the definition of $u_r(M)$ is attained, that is, so that

$$\min_j m_{kj} = u_r(M).$$

The amount actually won by playing strategy k might even be larger if the column player chooses a strategy unwisely.

A similar interpretation holds for the column player. Her best guaranteed payoff is obtained by playing a column l such that

$$\max_i m_{il} = v_c(M).$$

In the three examples just given, the values to the row player are, respectively, $-1, 1$, and 1 and the values to the column player are, respectively, $-1, 1$, and 2. The fact that the row player's value equals the column player's value in the two examples where a saddle point exists (and not in the other example) is no coincidence. Before proving this, we need a lemma.

LEMMA 2.2. *For any matrix M,*

$$u_r(M) \leq u_c(M).$$

PROOF. For any i and any l, we have

$$\min_j m_{ij} \leq m_{il}.$$

Maximizing both sides over i gives

$$u_r(M) = \max_i \min_j m_{ij} \leq \max_i m_{il},$$

for every l. Hence

$$u_r(M) \leq \min_l \max_i m_{il} = u_c(M). \qquad \square$$

THEOREM 2.3. *If the matrix game M has a saddle point m_{pq}, then*

$$u_r(M) = u_c(M) = m_{pq}.$$

PROOF. We have that

$$\min_j m_{pj} = m_{pq},$$

and so

$$u_r(M) \geq m_{pq}.$$

Also

$$\max_i m_{iq} = m_{pq},$$

and so

$$u_c(M) \leq m_{pq}.$$

Combining these, we have

$$u_c(M) \leq m_{pq} \leq u_r(M).$$

But, from the lemma,

$$u_r(M) \leq u_c(M),$$

and so the two values are equal. $\square$

The converse also holds.

THEOREM 2.4. *If $u_r(M) = u_c(M)$, then M has a saddle point.*

PROOF. Choose k such that

$$\min_j m_{kj} = u_r(M).$$

Then choose l such that

$$m_{kl} = \min_j m_{kj} = u_r(M) = u_c(M).$$

Now m_{kl} is a minimum in row k. There exists a column q such that

$$\max_i m_{iq} = u_c(M).$$

Thus

$$m_{kl} = u_c(M) = \max_i m_{iq} \geq m_{kq}.$$

Since m_{kl} is a minimum in its row, we have

$$m_{kl} = m_{kq},$$

and so m_{kq} is also a minimum in its row. Finally,

$$m_{kq} = m_{kl} = \max_i m_{iq},$$

and so m_{kq} is a saddle point. $\square$

Let M be a matrix with a saddle point. Suppose that the row player and column player play row i and column j, respectively. If row i contains a saddle point, say m_{il}, then

$$m_{ij} \geq m_{il} = u_r(M).$$

On the other hand, if row i does not contain a saddle point, while column j does contain one (say, m_{kj}), then

$$m_{ij} \leq m_{kj} = u_r(M).$$

In other words, the row player, by playing a row containing a saddle point, can guarantee himself a payoff of at least $u_r(M)$. The consequence of not playing such a row is that the column player could (and would) play so as to make the row player's payoff at most $u_r(M)$. From these considerations, it is clear that the best play for the row player is a row containing a saddle point. By similar reasoning, the column player should play a column containing a saddle point. If they both follow our recommendation, then, by Theorem 2.1, the payoff entry m_{ij} will be a saddle point.

In summary, if a matrix game M has a saddle point, then its *solution* is:

- The row player plays any row containing a saddle point.
- The column player plays any column containing a saddle point.
- The payoff to the row player is $u_r(M) = u_c(M)$, while the payoff to the column player is, of course, $-u_r(M) = -u_c(M)$.

Exercises

(1) Find all saddle points of the matrix

$$\begin{pmatrix} 2 & -1 & 0 & 0 & 1 \\ 0 & 0 & 1 & 2 & 1 \\ 1 & -2 & 1 & 0 & 2 \\ -1 & 0 & -1 & 1 & 1 \end{pmatrix}.$$

(2) Find all saddle points of the matrix

$$\begin{pmatrix} 1 & -1 & 2 & 2 & 0 \\ -2 & 0 & 1 & 0 & 2 \\ -1 & 0 & -1 & -1 & 0 \\ 1 & 1 & 1 & 2 & 1 \\ 1 & -1 & 0 & -1 & 1 \end{pmatrix}.$$

(3) For which values of a does the following matrix have a saddle point:

$$\begin{pmatrix} -2 & a \\ a & 1 \end{pmatrix}?$$

(4) For which values of a does the following matrix have a saddle point:

$$\begin{pmatrix} 1 & a \\ 2 & -1 \end{pmatrix}?$$

(5) For the following matrix, compute $u_r(M)$ and $u_c(M)$:

$$M = \begin{pmatrix} 0 & 1 & 1 & 2 \\ 1 & -1 & 3 & 1 \\ 2 & 0 & 0 & 2 \\ 3 & 2 & 1 & -1 \end{pmatrix}.$$

(6) For the following matrix, compute $u_r(M)$ and $u_c(M)$:

$$M = \begin{pmatrix} 1 & -1 & 2 & 0 & 1 \\ 2 & 0 & -1 & 1 & -1 \\ -2 & 2 & 0 & 0 & 1 \\ 0 & 1 & -1 & 0 & 2 \end{pmatrix}.$$

2.2. Mixed Strategies

The results of the previous section show that matrix games with saddle points can be easily solved. Let us consider a matrix without any saddle points. Here is a 2×2 example:

$$M = \begin{pmatrix} 2 & -1 \\ -1 & 1 \end{pmatrix}.$$

It is easy to see that if the row player consistently plays strategy 1, then the column player, when she notices this, will play 2 consistently and win 1 every time. On the other hand, if the row player plays 2 consistently, then the column player will play 1 and again win 1. The same result would ensue if the row player varied his strategy, but in a predictable way. For example, if the row player decided to play 1 on odd-numbered days and major holidays, and 2 at other times, then the column player would eventually catch on to this and respond appropriately. In case it is the column player who adopts a fixed strategy or a predictable pattern of strategies, the row player could always respond so as to win either 1 or 2. A third possibility is that both players play flexibly, each responding to the other's previous moves. In this case, the row player might begin by playing 1. When the column player catches on, she starts playing 2. Then the row player switches to 2, whereupon the column player goes to 1. Then the row player goes back to 1. This rather mindless cycle could be repeated forever.

A new idea is needed in order to get any closer to a satisfactory concept of a solution to a matrix game: Each player should choose, at each play of the game, a strategy *at random*. In this way, the other player has no way of predicting which strategy will be used. The *probabilities* with which the various strategies are chosen will probably be known to the opponent, but the particular strategy chosen at a particular play of the game will not be known. The problem for each player will then be to set these probabilities in an optimal way.

Thus, we have the following:

DEFINITION 2.3. Let M be an $m \times n$ matrix game. A *mixed strategy* for the row player is an m-tuple $\vec{p}$ of probabilities. That is,

$$p_i \geq 0, \qquad 1 \leq i \leq m,$$

and

$$\sum_{i=1}^{m} p_i = 1.$$

Similarly, a mixed strategy for the column player is an n-tuple $\vec{q}$ of probabilities. That is,

$$q_j \geq 0, \qquad 1 \leq j \leq n,$$

and

$$\sum_{j=1}^{n} q_j = 1.$$

The idea of a mixed strategy for the row player is that he will, at each play of the game, choose his strategy at random, and that this choice will be made so that the probability of choosing strategy i is p_i. Thus, there is no way for the column player to predict which particular strategy will be played against her.

In practice, this choice of strategy can be carried out by any convenient "chance device." For instance, suppose that the row player in the 2×2 game which we just discussed decides to play according to $\vec{p} = (1/2, 1/2)$. Then he could simply flip a fair coin every time the game is played (without letting his opponent see how it comes up). He could then play row 1 if the toss is heads and row 2 if it is tails. For the same game, if the column player wants to use the mixed strategy $\vec{q} = (1/3, 2/3)$, she could achieve this by glancing at a watch with a second hand just before playing. If it shows between 0 and 20 seconds, she would play column 1, otherwise column 2. Less homely chance devices would be needed in more complicated situations. A random-number generator (found on most calculators) can always be used.

To emphasize the distinction between mixed strategies and ordinary strategies, we will refer to the latter as *pure strategies*. It should, however, be realized that pure strategies are really special cases of mixed strategies. The mixed strategy $\vec{p}$ for which $p_i = 0$ for $i \neq k$ (and $p_k = 1$) is obviously the same as the pure strategy k.

If one or both players adopt mixed strategies, the payoffs at each play of the game depend on which particular pure strategies happened to be chosen. The important quantity to use in studying mixed strategies is the *expected payoff*. It is an average over many plays of the game and is denoted $E(\vec{p}, \vec{q})$, where $\vec{p}$ and $\vec{q}$ are the mixed strategies in use. To see how to define this quantity, consider first the case where the row player plays according to the mixed strategy $\vec{p}$ and the column player plays the pure strategy j. Then, for an $m \times n$ matrix M, the payoff to the row player is m_{ij} with probability p_i. The expected payoff is thus

$$E(\vec{p}, j) = \sum_{i=1}^{m} p_i m_{ij}.$$

Now, if the column player adopts the mixed strategy $\vec{q}$, the payoff to the row player is $E(\vec{p}, j)$ with probability q_j. Therefore

$$E(\vec{p}, \vec{q}) = \sum_{j=1}^{n} q_j \sum_{i=1}^{m} p_i m_{ij} = \sum_{j=1}^{n} \sum_{i=1}^{m} p_i q_j m_{ij}.$$

Note also that, interchanging the order of summation, we get

$$E(\vec{p}, \vec{q}) = \sum_{i=1}^{m} p_i \sum_{j=1}^{n} q_j m_{ij}.$$

Let us compute some expected payoffs. With M the 2×2 matrix presented at the beginning of this section, suppose the row player plays the mixed strategy $\vec{p} = (1/2, 1/2)$ and the column player plays the mixed strategy $\vec{q} = (2/3, 1/3)$. Then

$$E((1/2, 1/2), (2/3, 1/3)) =$$
$$2(2/3)(1/2) - (1/3)(1/2) - (2/3)(1/2) + (1/3)(1/2) = 1/3.$$

Now suppose that the column player plays $\vec{q} = (2/3, 1/3)$ as above. Given that information, how should the row player play so as to maximize his expected payoff? To compute his best strategy, note first that every mixed strategy for him is of the form $(p, 1 - p)$, where $0 \le p \le 1$. Now

$$E((p, 1 - p), (2/3, 1/3)) = 4p/3 - 1/3.$$

From this we see that the maximum expected payoff is 1 and is attained for $p = 1$. That is, the row player should play the pure strategy 1.

2.2.1. Row Values and Column Values

We now make a pair of definitions which are analogous to the definitions of row player's and column player's values which we made earlier. The change is that mixed strategies are substituted for pure strategies.

DEFINITION 2.4. Let M be an $m \times n$ matrix game. The *row value* is defined

$$v_r(M) = \max_{\vec{p}} \min_{\vec{q}} E(\vec{p}, \vec{q}),$$

where $\vec{p}$ and $\vec{q}$ range over all mixed strategies for the row player and column player, respectively. Similarly, the *column value* is defined

$$v_c(M) = \min_{\vec{q}} \max_{\vec{p}} E(\vec{p}, \vec{q}).$$

Thus $v_r(M)$ is an amount that the row player is *guaranteed* to win on the average, assuming that he plays intelligently. He *may* win more if the column player makes mistakes, but he cannot count on it. It will be proved in a later chapter that the maximum in the definition of v_r is actually attained, that is, that there exists at least one mixed strategy $\vec{r}$ for the row player such that

$$v_r(M) = \min_{\vec{q}} E(\vec{r}, \vec{q}).$$

Such a strategy is called an *optimal mixed strategy* for the row player. Similarly, there exists at least one optimal mixed strategy $\vec{s}$ for the column player such that

$$v_c(M) = \max_{\vec{p}} E(\vec{p}, \vec{s}).$$

There is an important point to be made about these optimal mixed strategies. If the row player plays such a strategy, he is guaranteed to gain at least $v_r(M)$ even if the column player plays as well as she can. On the other hand, if the row player knows that the column player is playing a stupid strategy, then there may well be a counter-strategy which gains him more than the "optimal" one does.

For now, we will simply assume that we already know that these optimal mixed strategies exist. Another theorem (called the minimax theorem), which will be proved later, is that the row and column values are always equal. This is a surprising and vital piece of information. First, it is surprising because it is not at all clear from the definitions why it should be true. Second, it is vital because the whole theory we are developing in this chapter would not work without it. To gain some insight into this matter, we prove the following:

THEOREM 2.5. *Let M be an $m \times n$ game matrix and let $\vec{r}$ and $\vec{s}$ be optimal mixed strategies for the row player and column player, respectively. Then*

$$v_r(M) \leq E(\vec{r}, \vec{s}) \leq v_c(M).$$

PROOF. Since $\vec{s}$ is a mixed strategy for the column player, we certainly have

$$v_r(M) = \min_{\vec{q}} E(\vec{r}, \vec{q}) \leq E(\vec{r}, \vec{s}).$$

Similarly, we have

$$v_c(M) = \max_{\vec{p}} E(\vec{p}, \vec{s}) \geq E(\vec{r}, \vec{s}).$$

Combining these two inequalities gives the result.   □

This theorem gives us the inequality

$$v_r(M) \leq v_c(M),$$

which is a weak version of the minimax theorem. More importantly, if we assume the minimax theorem, $v_r(M) = v_c(M)$, we get

$$v_r(M) = E(\vec{r}, \vec{s}) = v_c(M).$$

Now suppose, in addition, that $\vec{r}$ and $\vec{s}$ are optimal mixed strategies for the row player and column player, respectively. Then, if $\vec{p}$ and $\vec{q}$ are *any* mixed strategies for the players,

$$E(\vec{p}, \vec{s}) \leq v_c(M) = E(\vec{r}, \vec{s}) = v_r(M) \tag{2.1}$$

and

$$E(\vec{r}, \vec{q}) \geq v_r(M) = E(\vec{r}, \vec{s}) = v_c(M). \tag{2.2}$$

Then (2.1) and (2.2) together say that the row player can do no worse than $v_r(M)$ by playing $\vec{r}$, and may do worse than $v_r(M)$ playing $\vec{p}$. Thus, the row player *should* play an optimal mixed strategy. Similarly, the column player *should* play an optimal mixed strategy.

We have the following:

DEFINITION 2.5. Let M be a matrix game for which $v_r(M) = v_c(M)$. A *solution* to M consists of three components:

- An optimal mixed strategy for the row player.
- An optimal mixed strategy for the column player.
- The *value* of the game, $v(M)$, defined by

$$v(M) = v_r(M) = v_c(M).$$

Inequalities (2.1) and (2.2) are reminiscent of the definition of a saddle point in a matrix. In fact, if we think of the mixed strategies as labeling the rows and columns of the infinite matrix whose entries are the expected payoffs $E(\vec{p}, \vec{q})$, then (2.1) and (2.2) say that the pair $(\vec{r}, \vec{s})$ of optimal mixed strategies is a saddle point. The following theorem is analogous to Theorem 2.1.

THEOREM 2.6. *Let M be a matrix game for which $v_r(M) = v_c(M)$. Suppose that $\vec{r}$ and $\vec{t}$ are both optimal mixed strategies for the row player, while $\vec{s}$ and $\vec{u}$ are both optimal mixed strategies for the column player. Then*

$$E(\vec{r}, \vec{s}) = E(\vec{r}, \vec{u}) = E(\vec{t}, \vec{u}) = E(\vec{t}, \vec{s}).$$

PROOF. Using (2.1) and (2.2), we have

$$E(\vec{r}, \vec{s}) \geq E(\vec{t}, \vec{s}) \geq E(\vec{t}, \vec{u}),$$

and

$$E(\vec{t}, \vec{u}) \geq E(\vec{r}, \vec{u}) \geq E(\vec{r}, \vec{s}).$$

Thus, all the inequalities above are really equalities, and the proof follows. □

It follows that, if there is a choice, it does not matter which optimal mixed strategies the players choose.

Since saddle points of game matrices correspond to equilibrium pairs of pure strategies, we see that a pair of optimal mixed strategies is analogous to an equilibrium pair of mixed strategies. This concept will be formally defined and discussed in detail in Chapter 5.

We emphasize that once the minimax theorem has been proved we will know that *every* matrix M satisfies the condition $v_r(M) = v_c(M)$ and thus has a solution.

The following theorem gives us a somewhat simpler way to compute the row and column values. It allows us to use only pure strategies in computing the inside minimum (or maximum).

THEOREM 2.7. *Let M be an $m \times n$ matrix. Then*

$$v_r(M) = \max_{\vec{p}} \min_{j} E(\vec{p}, j)$$

and

$$v_c(M) = \min_{\vec{q}} \max_{i} E(i, \vec{q}).$$

Here, j ranges over all columns and i ranges over all rows.

PROOF. Since pure strategies are special cases of mixed strategies, we certainly have

$$\min_{\vec{q}} E(\vec{p}, \vec{q}) \le \min_{j} E(\vec{p}, j),$$

for any mixed strategy $\vec{p}$ for the row player. To prove the opposite inequality, let l be such that

$$E(\vec{p}, l) = \min_{j} E(\vec{p}, j).$$

Then, if $\vec{q}$ is a mixed strategy for the column player, we have

$$E(\vec{p}, \vec{q}) = \sum_{j=1}^{n} q_j E(\vec{p}, j) \ge E(\vec{p}, l).$$

Since this holds for all $\vec{q}$, we have, by the choice of l,

$$\min_{\vec{q}} E(\vec{p}, \vec{q}) \ge \min_{j} E(\vec{p}, j).$$

Thus, for all $\vec{p}$, we have

$$\min_{\vec{q}} E(\vec{p}, \vec{q}) = \min_{j} E(\vec{p}, j).$$

Maximizing over $\vec{p}$ on both sides of this equation completes the proof for the row value. The proof for the column value is similar. $\square$

This proof also shows that a strategy $\vec{r}$ for the row player is optimal if and only if

$$v_r(M) = \min_j E(\vec{r}, j),$$

where j runs over all columns. Similarly, a strategy $\vec{s}$ for the column player is optimal if and only if

$$v_c(M) = \max_i E(i, \vec{s}),$$

where i runs over all rows.

This theorem will play an important role in the remaining sections of this chapter.

If M has a saddle point, say m_{kl}, then the pure strategies k and l for the row player and column player, respectively, together with the value, m_{kl}, constitute a solution. This follows from:

COROLLARY 2.8. *If M has a saddle point m_{kl}, then*

$$v_r = v_c = m_{kl},$$

and so k and l are optimal mixed strategies.

PROOF. Since pure strategies are special cases of mixed strategies, we see from the theorem that

$$v_r(M) = \max_{\vec{p}} \min_j E(\vec{p}, j) \ge u_r(M).$$

Similarly,

$$v_c(M) \le u_c(M).$$

Since $u_r(M) = u_c(M)$ and $v_r(M) \le v_c(M)$, we see that

$$v_r(M) = v_c(M) = u_r(M) = u_c(M) = m_{kl}.$$

To show that k is optimal for the row player, notice that

$$\min_j E(k, j) = \min_j m_{kj} = m_{kl} = v_r(M).$$

By the remark made after the proof of the theorem, k is optimal. Similarly, l is optimal. □

2.2.2. *Dominated Rows and Columns*

It is sometimes possible to assign probability zero to some of the pure strategies. In other words, certain pure strategies can sometimes be identified as ones which would never appear with positive probability in an optimal strategy.

DEFINITION 2.6. Let M be an $m \times n$ matrix. Then row i *dominates* row k if

$$m_{ij} \geq m_{kj} \text{ for all } j.$$

Also, column j *dominates* column l if

$$m_{ij} \leq m_{il} \text{ for all } i.$$

Note that the inequality in the definition of domination of columns is reversed compared to the inequality in domination of rows. It should be obvious that a dominated row need never be used by the row player, and that a dominated column need never be used by the column player. This implies that if we are looking for optimal mixed strategies, we may as well assign probabilities of zero to any such row or column. In fact, we may as well reduce the size of the matrix by erasing dominated rows and columns. Let us consider an example. Let

$$M = \begin{pmatrix} 1 & -1 & -2 \\ 2 & -1 & 0 \\ -1 & 1 & 1 \end{pmatrix}.$$

We see that row 1 is dominated by row 2. We erase the dominated row to get

$$\begin{pmatrix} 2 & -1 & 0 \\ -1 & 1 & 1 \end{pmatrix}.$$

In this new matrix, column 3 is dominated by column 2 (there was no domination of columns in the original matrix). We arrive at

$$\begin{pmatrix} 2 & -1 \\ -1 & 1 \end{pmatrix},$$

in which there is no domination.

We have the following:

DEFINITION 2.7. Let M be a matrix game and let $\vec{p}$ be a mixed strategy for the row player. Then row i of M is said to be *active* in $\vec{p}$ if $p_i > 0$. Similarly, if $\vec{q}$ is a mixed strategy for the column player, then column j is *active* in $\vec{q}$ if $q_j > 0$. A row or column which is not active is said to be *inactive*.

Our discussion of domination can be summarized: A dominated row (or column) is inactive in an optimal strategy for the row player (or column player). [But see Exercise (13).]

The following theorem will be useful.

THEOREM 2.9. *Let M be an $m \times n$ matrix game such that $v_r(M) = v_c(M)$. Let $\vec{r}$ and $\vec{s}$ be mixed strategies for the row player and column player, respectively. Then $\vec{r}$ and $\vec{s}$ are both optimal if and only if the following two conditions hold:*

(1) *Row k is inactive in $\vec{r}$ whenever*

$$E(k, \vec{s}) < \max_i E(i, \vec{s}).$$

(2) *Column l is inactive in $\vec{s}$ whenever*

$$E(\vec{r}, l) > \min_j E(\vec{r}, j).$$

PROOF. Suppose first that $\vec{r}$ and $\vec{s}$ are optimal for the row and column player, respectively. To show that (1) holds, suppose that

$$E(k, \vec{s}) < \max_i E(i, \vec{s}) = v_c(M),$$

but that $r_k > 0$. Then

$$E(\vec{r}, \vec{s}) = \sum_{i=1}^{m} r_i E(i, \vec{s}) < \sum_{i=1}^{m} r_i v_c(M) = v_c(M),$$

since

$$E(i, \vec{s}) \le v_c(M) \quad \text{for all } i.$$

This contradicts the optimality of $\vec{r}$ and $\vec{s}$ (see the remarks after Theorem 2.5). The proof that (2) holds is similar.

Now suppose that both (1) and (2) hold. We must show that $\vec{r}$ and $\vec{s}$ are optimal. Using (1) and (2), together with Theorem 2.7, we have

$$E(\vec{r}, \vec{s}) = \sum_{i=1}^{m} r_i E(i, \vec{s}) = \max_i E(i, \vec{s}) \ge v_c(M),$$

and

$$E(\vec{r}, \vec{s}) = \sum_{j=1}^{n} s_j E(\vec{r}, j) = \min_j E(\vec{r}, j) \le v_r(M).$$

Since $v_r(M) = v_c(M)$, we have

$$v_c(M) = E(\vec{r}, \vec{s}) = \max_i E(i, \vec{s}),$$

and

$$v_r(M) = E(\vec{r}, \vec{s}) = \min_j E(\vec{r}, j).$$

Thus, both $\vec{r}$ and $\vec{s}$ are optimal. □

This theorem is occasionally useful in computing optimal strategies. In fact, it will play such a role in the next section. It is very useful for *verifying* optimality of strategies. For example, let

$$M = \begin{pmatrix} 2 & -1 & -1 \\ -2 & 0 & 3 \\ 1 & 2 & 1 \end{pmatrix}.$$

We use Theorem 2.9 to verify that $\vec{r} = (0,0,1)$ and $\vec{s} = (2/3,0,1/3)$ are optimal strategies for the row player and column player, respectively. We have

$$E(1,\vec{s}) = 1 \quad E(2,\vec{s}) = -1/3 \quad E(3,\vec{s}) = 1,$$

and

$$E(\vec{r},1) = 1 \quad E(\vec{r},2) = 2 \quad E(\vec{r},3) = 1.$$

Thus

$$\max_i E(i,\vec{s}) = 1 \quad \text{and} \quad \min_j E(\vec{r},j) = 1.$$

Theorem 2.9 says that, in order for both $\vec{r}$ and $\vec{s}$ to be optimal, we must have r_2 and s_2 equal zero. They both are, and so optimality is verified.

Exercises

(1) Let

$$M = \begin{pmatrix} 1 & 2 & 3 \\ 3 & 0 & 2 \\ 2 & 1 & 0 \end{pmatrix}.$$

If the column player plays the mixed strategy $(2/5, 1/3, 4/15)$, what is the best way for the row player to play?

(2) Let

$$M = \begin{pmatrix} 1 & -1 & 1 \\ -1 & 0 & 1 \\ 2 & -1 & 0 \end{pmatrix}.$$

(a) Compute $E((1/5,2/5,2/5),(1/3,1/3,1/3))$. (b) On the assumption that the row player continues to play $(1/5,2/5,2/5)$, what is the best way for the column player to play?

(3) Let

$$M = \begin{pmatrix} 3 & 2 & -1 & 0 \\ 1 & 1 & -1 & -1 \\ 0 & -1 & 1 & 2 \end{pmatrix}.$$

Eliminate dominated rows and columns so as to reduce M to the smallest size possible.

(4) Suppose you are the row player in a 3×4 game and want to play the mixed strategy $\vec{p} = (5/12, 1/4, 1/3)$. Your calculator can produce 8-digit random numbers uniformly distributed in the interval $[0,1]$. Explain how you would use these random numbers to achieve $\vec{p}$ (as closely as possible).

(5) For the matrix game

$$\begin{pmatrix} -1 & 2 & -2 & 0 & 1 \\ -2 & -1 & 3 & 2 & 0 \\ 2 & 1 & 0 & -1 & -2 \\ 0 & 0 & 2 & 1 & 1 \\ 1 & -1 & 0 & -2 & 1 \end{pmatrix},$$

verify that

$$\vec{p} = (5/52, 0, 11/52, 17/26, 1/26), \vec{q} = (21/52, 3/13, 0, 3/52, 4/13), v = 19/52,$$

is a solution.

(6) Given

$$\begin{pmatrix} 1 & -1 & 2 \\ -2 & 0 & 1 \\ 0 & 2 & -1 \end{pmatrix},$$

verify that the following is a solution:

$$\vec{p} = (1/2, 0, 1/2), \vec{q} = (3/4, 1/4, 0), v(M) = 1/2.$$

(7) Prove that if $\vec{p}$ and $\vec{q}$ are mixed strategies for the row player and column player, respectively, such that

$$\min_{j} E(\vec{p}, j) = \max_{i} E(i, \vec{q})$$

then $\vec{p}$ and $\vec{q}$ are optimal.

(8) Given

$$\begin{pmatrix} 2 & -3 & 4 & -5 \\ -1 & 2 & -3 & 4 \\ 0 & 1 & -2 & 3 \\ 1 & 2 & -3 & 4 \\ -3 & 4 & -5 & 6 \end{pmatrix},$$

verify that $\vec{p} = (1/2, 0, 1/6, 0, 1/3)$ and $\vec{q} = (0, 1/4, 1/2, 1/4)$ are optimal strategies.

(9) Let M be a 2×2 matrix. Prove that M has a saddle point if and only if it has a dominated row or a dominated column. Is this true for larger matrices?

(10) Suppose that both $\vec{r}$ and $\vec{u}$ are optimal strategies for the row player for the matrix game M. Prove that if $0 \le t \le 1$ then $t\vec{r} + (1 - t)\vec{u}$ is also an optimal strategy for the row player.

(11) Let M be a matrix and c a real number. Form the matrix M' by adding c to every entry of M. Prove that

$$v_r(M') = v_r(M) + c \quad \text{and} \quad v_c(M') = v_c(M) + c.$$

(12) Let M be a matrix and a a nonnegative real number. Prove that

$$v_r(aM) = av_r(M) \quad \text{and} \quad v_c(aM) = av_c(M).$$

What happens if $a < 0$?

(13) The assertion was made that a dominated row or column is never active in an optimal strategy. Explain why this statement may be incorrect if two of the rows or columns are equal.

2.3. Small Games

In this section, we discuss a graphical method for solving those matrix games in which at least one of the players has only two pure strategies. We start with the following 2×2 example:

$$M = \begin{pmatrix} 2 & -3 \\ -1 & 1 \end{pmatrix}.$$

We know, from Theorem 2.7, that

$$v_r(M) = \max_{\vec{p}} \min_j E(\vec{p}, j).$$

Now each $\vec{p}$ is of the form $(p, 1-p)$, where $0 \le p \le 1$. For convenience of notation, we write

$$\pi_j(p) = E((p, 1-p), j) \quad \text{for } j = 1, 2 \text{ and } 0 \le p \le 1.$$

These functions of p are easy to compute. We get

$$\pi_1(p) = 2p - (1-p) = 3p - 1$$

and

$$\pi_2(p) = -3p + (1-p) = -4p + 1.$$

Now these two functions are both linear. In Figure 2.1, we graph both of them and indicate, with a heavier line, their minimum. By definition, $v_r(M)$ is the maximum of this minimum. It is circled in the figure. It occurs at the point where $\pi_1(p)$ crosses $\pi_2(p)$. Setting $\pi_1(p) = \pi_2(p)$, we get that $v_r(M) = -1/7$ and it is attained for $p = 2/7$.

Thus $\vec{p} = (2/7, 5/7)$ is an optimal mixed strategy for the row player. To find an optimal strategy for the column player, let us define

$$\pi^i(q) = E(i, (q, 1-q)) \quad \text{for } i = 1, 2 \text{ and } 0 \le q \le 1.$$

Thus, $v_c(M)$ is the minimum over q of the maximum of the two linear functions $\pi^1(q)$ and $\pi^2(q)$. In Figure 2.2, we graph both of them. Their

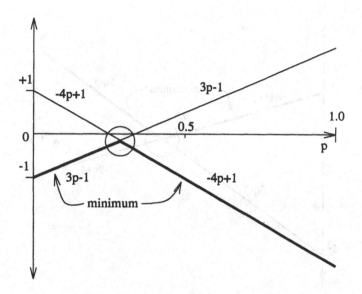

FIGURE 2.1. Finding the row value.

maximum is indicated with a heavier line and the minimum of this maximum is circled. We have

$$\pi^1(q) = 2q - 3(1 - q) = 5q - 3$$

and

$$\pi^2(q) = -q + (1 - q) = -2q + 1.$$

The minimum of the maximum occurs where these cross and so we easily
compute $v_c(M) = -1/7$, attained at $\vec{q} = (4/7, 3/7)$. Since $v_r(M) = v_c(M)$,
we have indeed solved the game.

To solve a 2×2 matrix game, it is not really necessary to draw the
graph. This follows from:

THEOREM 2.10. *Let*

$$M = \begin{pmatrix} a & b \\ c & d \end{pmatrix}$$

*be a 2×2 matrix and suppose M has no saddle points. Then, (i) the straight
lines $\pi_1(p)$ and $\pi_2(p)$ cross at a value p^* of p in the open interval $(0,1)$;
moreover,*

$$\pi_1(p^*) = \pi_2(p^*) = v_r(M).$$

*Also, (ii) the straight lines $\pi^1(q)$ and $\pi^2(q)$ cross at a value q^* in the open
interval $(0,1)$; moreover,*

$$\pi^1(q^*) = \pi^2(q^*) = v_c(M).$$

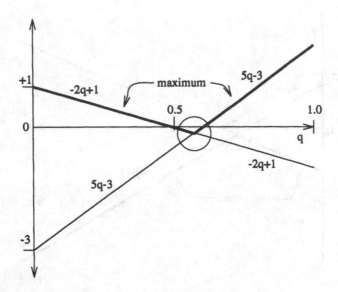

FIGURE 2.2. Finding the column value.

PROOF. We prove (i); the proof of (ii) is similar.

We must first prove that the two lines cross in the open interval $(0, 1)$. Suppose that this is not true. Then one of the two functions must lie above the other. That is, we must have either

$$\pi_1(p) \geq \pi_2(p), \quad 0 \leq p \leq 1, \qquad (2.3)$$

or

$$\pi_2(p) \geq \pi_1(p), \quad 0 \leq p \leq 1. \qquad (2.4)$$

If the first of these inequalities occurs, then we have that

$$c = \pi_1(0) \geq \pi_2(0) = d \quad \text{and} \quad a = \pi_1(1) \geq \pi_2(1) = b.$$

Thus, column 1 is dominated by column 2. By Exercise (9) of Section 2.2, M has a saddle point. This is a contradiction. If (2.4) holds, then column 2 is dominated by column 1, which is again a contradiction.

We have verified that the two lines cross in $(0, 1)$. Let p^* denote the value of p where this happens. Now we show that the maximum of the minimum occurs at p^*. To the left of p^*, one of the two functions is strictly less than the other. Let us assume that $\pi_1(p) < \pi_2(p)$ for $0 \leq p < p^*$ (the other case is similar). Thus, taking $p = 0$, we see that $c < d$. Consideration of the geometry of the situation shows that we only need to prove that $\pi_1(p)$ has positive slope and that $\pi_2(p)$ has negative slope. It is easily verified that the slope of $\pi_1(p)$ is $a - c$ and that the slope of $\pi_2(p)$ is $b - d$. First, suppose that $a - c \leq 0$. Then $a \leq c$. Thus, since $c < d$, c is a saddle point. This is a contradiction and so $a - c > 0$. Second, suppose that $b - d \geq 0$. Then, since $a > c$, row 1 dominates row 2. This means, as before, that M has a saddle point, which is a contradiction. $\square$

Let us look at another example. Consider

$$M = \begin{pmatrix} 1 & 0 & 1 \\ -1 & 0 & 1 \\ -1 & 2 & 1 \end{pmatrix}.$$

We see that there are no saddle points, but that row 2 is dominated. After removing it, we see that column 3 is dominated. Erasing it leaves us with

$$\begin{pmatrix} 1 & 0 \\ -1 & 2 \end{pmatrix}.$$

This has no saddle points and so Theorem 2.10 applies. Now

$$\pi_1(p) = 2p - 1 \quad \text{and} \quad \pi_2(p) = -2p + 2.$$

Setting these equal, we solve to get $p^* = 3/4$. The row value is thus $\pi_1(3/4) = \pi_2(3/4) = 1/2$. As for the column player, we see that

$$\pi^1(q) = q \quad \text{and} \quad \pi^2(q) = -3q + 2.$$

Setting these equal, we get $q^* = 1/2$. The column value is then $1/2$. For the original 3×3 game, the probabilities of the erased row and column are both 0. The probabilities of the remaining rows and columns are the numbers just computed. We get that the optimal mixed strategy for the row player is $(3/4, 0, 1/4)$ and the optimal strategy for the column player is $(1/2, 1/2, 0)$.

2.3.1. $2 \times n$ and $m \times 2$ Games

Now we consider $2 \times n$ matrix games. Here is an example:

$$M = \begin{pmatrix} 4 & -4 & 1 \\ -4 & 4 & -2 \end{pmatrix}.$$

There are no saddle points and no dominated rows or columns. We compute $\pi_j(p) = E((p, 1 - p), j)$ for $j = 1, 2, 3$ and $0 \leq p \leq 1$, and get

$$\pi_1(p) = 8p - 4, \quad \pi_2(p) = -8p + 4, \quad \pi_3(p) = 3p - 2.$$

Then we graph these three linear functions (in Figure 2.3), indicate their minimum by a heavier line, and circle the maximum of this minimum.

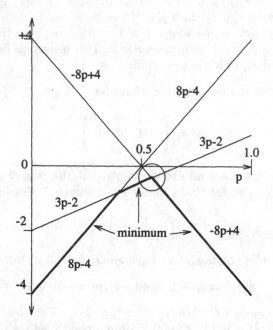

FIGURE 2.3. Row value for a 2×3 game

We see from the graph that the maximum of the minimum occurs where $\pi_2(p)$ crosses $\pi_3(p)$. Setting these equal, we get $p^* = 6/11$ and so $v_r(M) = \pi_2(6/11) = \pi_3(6/11) = -4/11$. To compute the optimal strategy for the column player, we note, from Figure 2.3, that

$$\pi_1(p^*) > \min_j \pi_j(p^*).$$

It follows from Theorem 2.9 that column 1 is inactive in any optimal strategy for the column player. In other words, we can assign probability zero to column 1. This leaves us with a 2×2 problem. Solving for the column player's strategy gives $\vec{q} = (0, 3/11, 8/11)$ as optimal and $v_c(M) = -4/11$.

Finally, we consider $m \times 2$ matrix games. Here is an example:

$$M = \begin{pmatrix} 1 & -1 \\ -1 & 1 \\ 0 & 1/2 \\ 1/2 & 0 \end{pmatrix}.$$

There are no saddle points and no dominated rows or columns. We

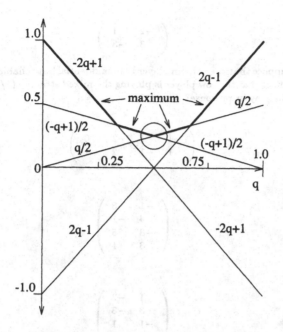

FIGURE 2.4. Column value for a 4×2 game.

compute $\pi^i(q) = E(i,(q,1-q))$ for $i = 1,2,3,4$ and $0 \leq q \leq 1$ and get

$$\pi^1(q) = 2q - 1, \ \pi^2(q) = -2q + 1, \ \pi^3(q) = -q/2 + 1/2, \ \pi^4(q) = q/2.$$

Then we graph these four linear functions (in Figure 2.4), indicate their maximum by a heavier line and circle the minimum of this maximum. We see that the minimum of the maximum occurs where $\pi^3(q)$ crosses $\pi^4(q)$. Setting these equal, we get $q^* = 1/2$ and so $v_c(M) = 1/4$. From the graph, we see that the first two rows will be inactive in an optimal strategy for the row player. This leaves us with a 2×2 problem. Solving for the row player's strategy, we get $\vec{p} = (0,0,1/2,1/2)$ and $v_r(M) = 1/4$.

Exercises

(1) First solve

$$\begin{pmatrix} -1 & 2 \\ 0 & -2 \end{pmatrix}.$$

Then suppose that the row player believes that the column player is playing the mixed strategy $(1/3, 2/3)$. Is there a strategy for the row player which is a better response to this than the "optimal" one? What is it?

(2) Solve

$$\begin{pmatrix} -2 & 1 \\ 3 & -3 \end{pmatrix}.$$

Then suppose that the column player has confidential but reliable information to the effect that the row player is playing the mixed strategy $(1/2, 1/2)$. How should he play in response?

(3) Solve

$$\begin{pmatrix} -1 & 1 & -2 & 0 \\ 1 & -1 & 2 & -1 \end{pmatrix}.$$

(4) Solve

$$\begin{pmatrix} -1 & 3 \\ 4 & -1 \\ -3 & 5 \\ 3 & 1 \end{pmatrix}.$$

(5) Solve

$$\begin{pmatrix} 1 & -2 \\ 3 & -3 \\ -1 & 1 \\ -2 & 3 \end{pmatrix}.$$

(6) For the matrix game

$$\begin{pmatrix} -1 & 1 & 1/2 \\ 1 & -1 & -1/2 \end{pmatrix},$$

compute the value, the optimal strategy for the row player, and *two* optimal strategies for the column player.

(7) How would you solve a $1 \times n$ or $m \times 1$ matrix game?

(8) Define a matrix game to be *fair* if its value is zero. Consider the matrix game

$$\begin{pmatrix} a & 2 \\ 1 & -1 \end{pmatrix}.$$

For which values of the parameter a is the game fair? When does it favor the row player (positive value)? When does it favor the column player?

(9) Let

$$M = \begin{pmatrix} -1 & a \\ 1 & 0 \end{pmatrix}.$$

For what values of a is this game fair [see Exercise (8)]? For what values of a does the row player have an advantage (that is, when is the value positive)? For what values of a does the column player have an advantage?

2.4. Symmetric Games

A game is *symmetric* if the players are indistinguishable except for their names. Thus, if the two players interchange their roles in such a game, neither would find it necessary to modify his or her optimal strategy. Two-finger morra is an example of a symmetric game. Chess is not symmetric since one of the two players moves first and the other second. For a matrix game M to be symmetric, it is first necessary that it be square (since the two players have the same number of pure strategies). Also, such a matrix must have the property that if the row player and column player interchange strategies, then they interchange payoffs. Thus the entry m_{ji} must be the negative of m_{ij}. This property is expressed in the following standard definition in linear algebra.

DEFINITION 2.8. A square matrix M of size $n \times n$ is *skew-symmetric* if

$$m_{ji} = -m_{ij}, \quad 1 \le i, j \le n.$$

Thus a *game* is symmetric if and only if its *matrix* is skew-symmetric. This collision in terminology between linear algebra and game theory is unfortunate but unavoidable. We will use the adjective "symmetric" to modify "game" and "skew-symmetric" to modify "matrix." An easy consequence of the definition is that every diagonal entry in a skew-symmetric matrix is zero (for, $m_{ii} = -m_{ii}$).

It is intuitively clear that neither player in a symmetric game has an advantage over the other. It follows that the row value must be nonpositive (otherwise, there would be a mixed strategy for the row player such that he would always win). By symmetry, the column value must be nonnegative. We give a formal proof of these facts.

THEOREM 2.11. *Let M be a symmetric matrix game. Then*

$$v_r(M) = -v_c(M)$$

and

$$v_r(M) \le 0 \le v_c(M).$$

PROOF. First, if $\vec{p}$ and $\vec{q}$ are any mixed strategies,

$$E(\vec{p}, \vec{q}) = \sum_{j=1}^{n}\sum_{i=1}^{n} p_i q_j m_{ij} = -\sum_{j=1}^{n}\sum_{i=1}^{n} p_i q_j m_{ji} = -E(\vec{q}, \vec{p}).$$

Then

$$v_r(M) = \max_{\vec{p}} \min_{\vec{q}} E(\vec{p}, \vec{q})$$

$$= \max_{\vec{p}} \min_{\vec{q}} (-E(\vec{q}, \vec{p}))$$

$$= \max_{\vec{p}} (-\max_{\vec{q}} E(\vec{q}, \vec{p}))$$

$$= -\min_{\vec{p}} \max_{\vec{q}} E(\vec{q}, \vec{p})$$

$$= -v_c(M).$$

Finally, by Theorem 2.5,

$$-v_c(M) = v_r(M) \le v_c(M).$$

Thus, $v_c(M) \ge 0$ and so $v_r(M) \le 0$. $\square$

COROLLARY 2.12. *If the symmetric matrix game M has a solution, then its value is zero. Also, if $\vec{r}$ is an optimal strategy for the row player, then $\vec{r}$ is also an optimal strategy for the column player (and vice versa).*

PROOF. If M has a solution, $v_r(M) = v_c(M)$. The fact that both these quantities are zero is immediate from the theorem.

Now

$$\max_{\vec{p}} E(\vec{p}, \vec{r}) = -\min_{\vec{p}} E(\vec{r}, \vec{p}) = -v_r(M) = 0,$$

and so $\vec{r}$ is optimal for the column player. $\square$

2.4.1. *Solving Symmetric Games*

We now present a method for solving symmetric matrix games. It has the advantage of being fairly easy to use if the game is small, and the disadvantage that it does not always work. The game of three-finger morra is an example for which the method fails. It is discussed in Chapter 4.

Observe that if $\vec{r}$ is an optimal strategy (for either player, and hence both) for an $n \times n$ symmetric matrix game M, then

$$E(\vec{r}, j) = \sum_{i=1}^{n} r_i m_{ij} \ge 0, \tag{2.5}$$

for all j, and that some of these inequalities must be equalities [otherwise $v_r(M)$ would be positive]. Now, treating the r_i's as unknowns, we have the equation

$$r_1 + \cdots + r_n = 1. \tag{2.6}$$

If we then choose $n-1$ of the inequalities (2.5) and set them to be equalities, we have a total of n equations. The idea then is to solve, if possible, this system of equations for the n unknowns to get a solution vector $\vec{r}$. If each component r_i of this solution vector is nonnegative, and if the remaining one of the inequalities (2.5) is valid for the r_i's, then we have found an optimal strategy. Consider an example:

$$M = \begin{pmatrix} 0 & -1 & 2 \\ 1 & 0 & -3 \\ -2 & 3 & 0 \end{pmatrix}.$$

This is clearly a skew-symmetric matrix. The inequalities (2.5) are, for $j = 1, 2, 3$, respectively,

$$r_2 - 2r_3 \geq 0,$$
$$-r_1 + 3r_3 \geq 0,$$
$$2r_1 - 3r_2 \geq 0.$$

Arbitrarily set the first two to be equations. These, together with (2.6), give us the system of equations

$$r_1 + r_2 + r_3 = 1,$$
$$r_2 - 2r_3 = 0,$$
$$-r_1 + 3r_3 = 0.$$

This system is easily solved to give: $r_1 = 1/2$, $r_2 = 1/3$, $r_3 = 1/6$. These are all nonnegative, and the third inequality (the one not used to solve for the r_i's) is valid. Thus we have solved the game. The probability vector $\vec{r} = (1/2, 1/3, 1/6)$ is an optimal strategy for both players and the value of the game is zero. The reader should verify that, in this case, setting any two of the three inequalities to zero gives this same solution.

Here is a second example:

$$M = \begin{pmatrix} 0 & -1 & 0 & -1 \\ 1 & 0 & -1 & 2 \\ 0 & 1 & 0 & -2 \\ 1 & -2 & 2 & 0 \end{pmatrix}.$$

The inequalities (2.5) are

$$r_2 + r_4 \geq 0,$$
$$-r_1 + r_3 - 2r_4 \geq 0,$$
$$-r_2 + 2r_4 \geq 0,$$
$$-r_1 + 2r_2 - 2r_3 \geq 0.$$

Before jumping into a calculation, let us think a bit. The first inequality has no negative coefficients. Thus, since a valid solution requires nonnegativity of the r_i's, setting it to be an equality would imply that $r_2 = r_4 = 0$.

That, together with the fourth inequality, would imply that r_1 and r_3 are zero also. This is impossible since the r_i's must sum to one. Thus, the only possible way to proceed is to set the last three inequalities to equalities. These, together with (2.6), give us a system of four equations in four unknowns. Solving it gives $\vec{r} = (0, 2/5, 2/5, 1/5)$. These are all nonnegative and satisfy the inequality not used in solving for them.

For a third example, consider

$$M = \begin{pmatrix} 0 & -1 & 0 & 1 \\ 1 & 0 & -2 & 0 \\ 0 & 2 & 0 & -2 \\ -1 & 0 & 2 & 0 \end{pmatrix}.$$

Using any three of the four inequalities leads to the parameterized set of solutions:

$$r_1 = 2/3 - 4r_4/3, \quad r_2 = r_4, \quad r_3 = 1/3 - 2r_4/3, \quad r_4 = r_4.$$

Substituting values of r_4 leads to infinitely many valid optimal strategies. For example, $(2/3, 0, 1/3, 0)$ and $(0, 1/2, 0, 1/2)$ are both solutions. On the other hand, we also get $(2, -1, 1, -1)$, which is invalid.

EXAMPLE 2.1. The game of *scissors-paper-stone* is played as follows. Each of the two players simultaneously makes a gesture indicating one of the three objects in the name of the game (a closed fist for "stone" etc.). If they choose the same object, the game is a draw. Otherwise, the winner is decided by the rules: "Scissors cut paper, paper covers stone, stone breaks scissors." The payoff is $+1$ for a win and -1 for a loss.

If the three pure strategies are listed in the order given in the name of the game, then the payoff matrix is

$$\begin{pmatrix} 0 & 1 & -1 \\ -1 & 0 & 1 \\ 1 & -1 & 0 \end{pmatrix}.$$

The game is certainly symmetric. The solution is left for Exercise (3).

Exercises

(1) Solve

$$\begin{pmatrix} 0 & -1 & 2 & -1 \\ 1 & 0 & -1 & -1 \\ -2 & 1 & 0 & 1 \\ 1 & 1 & -1 & 0 \end{pmatrix}.$$

(2) Solve
$$\begin{pmatrix} 0 & 1 & -2 \\ -1 & 0 & 3 \\ 2 & -3 & 0 \end{pmatrix}.$$

(3) Solve scissors-paper-stone.

(4) Solve
$$\begin{pmatrix} 0 & 1 & 2 & -3 \\ -1 & 0 & 1 & 0 \\ -2 & -1 & 0 & 1 \\ 3 & 0 & -1 & 0 \\ 2 & 0 & -1 & -1 \end{pmatrix}.$$

(5) Solve
$$\begin{pmatrix} 0 & 2 & -1 & -2 \\ -2 & 0 & 3 & -1 \\ 1 & -3 & 0 & 1 \\ 2 & 1 & -1 & 0 \end{pmatrix}.$$

(6) Find at least two optimal mixed strategies for two-finger morra.

(7) The game of two-finger morrette is played just like two-finger morra except that the payoff to the winner is always +1. Prove that it has infinitely many optimal strategies.

3
Linear Programming

The subject of this chapter is the part of the field of linear programming which we will need later in the book. We will use it to solve matrix games. In addition, the theoretical aspects of linear programming developed here will, when applied to game theory, give us the minimax theorem and other theorems about matrix games.

There are many good books which discuss linear programming in greater detail. See, for example, [Chv83], [Str89], or [Lue84].

3.1. Primal and Dual Problems

A *linear programming problem* consists of a linear real-valued *objective function*

$$w(\vec{x}) = c_1 x_1 + c_2 x_2 + \cdots + c_n x_n + d,$$

of n variables, which is either to be maximized or minimized *subject to* a finite set of linear *constraints*. Each constraint has either the form

$$a_1 x_1 + a_2 x_2 + \cdots + a_n x_n \leq b,$$

or the form

$$a_1 x_1 + a_2 x_2 + \cdots + a_n x_n \geq b.$$

We will sometimes abbreviate the phrase *linear programming problem* by *LP problem*.

Thus, the problem is to maximize or minimize $w(\vec{x})$ over the set of all vectors $\vec{x}$ which satisfy all the constraints. A *solution* consists of this maximum or minimum value, together with a vector $\vec{x}$ at which it is attained. In applications, the objective function often represents a quantity like profit or cost, while the variables represent quantities of various commodities. Here is a simple example of a linear programming problem:

$$\text{maximize} \quad -3x_1 + 2x_2 + x_3 \qquad (3.1)$$
$$\text{subject to} \quad x_1, x_2, x_3 \geq 0$$
$$x_1 + x_2 + x_3 \leq 1.$$

It is a maximization problem with three unknowns and a total of four constraints. The constraints

$$x_1, x_2, x_3 \geq 0$$

are called *positivity constraints*. In the great majority of all linear programming problems (including those arising from game theory), all the unknowns are naturally nonnegative. For this reason, we restrict ourselves to such problems; positivity constraints will henceforth be understood and not written. Thus, the problem just stated becomes

$$\text{maximize} \quad -3x_1 + 2x_2 + x_3 \qquad (3.2)$$
$$\text{subject to} \quad x_1 + x_2 + x_3 \leq 1.$$

We mention also that there is no loss of generality in restricting ourselves to problems in which all unknowns are constrained to be nonnegative. If a problem contains an unknown which is allowed to be negative, then we can replace it throughout by the difference of two new unknowns which *are* constrained to be nonnegative.

It should also be mentioned that *equality* constraints frequently arise. Such a constraint has the form

$$a_1 x_1 + a_2 x_2 + \cdots + a_n x_n = b.$$

This one constraint can be replaced by the equivalent pair of inequality constraints

$$a_1 x_1 + \cdots + a_n x_n \leq b,$$

and

$$a_1 x_1 + \cdots + a_n x_n \geq b.$$

Thus, the class of linear programming problems we are considering is really very general.

A vector $\vec{x}$ which satisfies all the constraints of a given linear programming problem is called a *feasible vector* for that problem. An LP problem is said to be *feasible* if there exists at least one feasible vector for it. Problem (3.2) is feasible since $\vec{x} = (0,0,0)$ is a feasible vector. On the other hand, the LP problem

$$\text{minimize} \quad 3x_1 + x_2 \tag{3.3}$$

$$\text{subject to} \quad x_1 + x_2 \leq -1$$

is infeasible because $x_1, x_2 \geq 0$ and so their sum cannot be negative. Note that the objective function is irrelevant for feasibility.

A feasible vector $\vec{x}$ is *optimal* if the objective function attains its maximum (or minimum) at $\vec{x}$. Some feasible problems have no solution because the objective function is unbounded. For example, the problem

$$\text{maximize} \quad w(\vec{x}) = -x_1 + 2x_2 \tag{3.4}$$

$$\text{subject to} \quad x_1 - x_2 \leq 2$$

is unbounded because the vector $\vec{x} = (0, x)$ is feasible for every $x \geq 0$ and

$$\lim_{x \to \infty} w((0, x)) = +\infty.$$

It will be proved later that every bounded feasible problem has a solution. If $w(\vec{x})$ is the objective function of a feasible maximization problem, we will use the notation $\max w(\vec{x})$ to denote its maximum value over the set of feasible vectors. If the problem is unbounded, we write $\max w(\vec{x}) = +\infty$. For a minimization problem, we use the corresponding notation $\min w(\vec{x})$. If the problem is unbounded, we write $\min w(\vec{x}) = -\infty$.

3.1.1. *Primal Problems and Their Duals*

The following definition gives an important special type of problem.

DEFINITION 3.1. A linear programming problem is said to be *primal* if it has the form

$$\text{maximize} \quad f(\vec{x}) = \sum_{j=1}^{n} c_j x_j + d$$

$$\text{subject to} \quad \sum_{j=1}^{n} a_{ij} x_j \leq b_i, \quad 1 \leq i \leq m,$$

where $A = (a_{ij})$ is an $m \times n$ coefficient matrix, $\vec{c}$ is an n-tuple of numbers, d is a constant, and $\vec{b}$ is an m-tuple of numbers.

Problems (3.2) and (3.4) are primal but Problem (3.3) is not. Here is another example.

$$\text{maximize} \quad f(x_1, x_2, x_3, x_4) = x_1 - x_4 \qquad (3.5)$$

$$\text{subject to} \quad x_1 + x_2 + x_3 - x_4 \le 2$$

$$-x_1 - 3x_2 - x_3 + 2x_4 \le -1.$$

It is feasible because, for example, $\vec{x} = (0, 0, 1, 0)$ is a feasible vector.

It should be noticed that *any* linear programming problem can be converted into an equivalent problem which is primal. The method for doing so is briefly stated as follows: If the objective function is to be minimized, replace it with its negative; if there is a constraint in which the left-hand side is greater than or equal to the right-hand side, take negatives on both sides.

A primal problem is really half of a pair of problems. The other half is called the *dual* problem. We have the following:

DEFINITION 3.2. Consider the primal problem

$$\text{maximize} \quad f(\vec{x}) = \sum_{j=1}^{n} c_j x_j + d$$

$$\text{subject to} \quad \sum_{j=1}^{n} a_{ij} x_j \le b_i, \quad 1 \le i \le m.$$

The *dual problem* corresponding to this primal problem is

$$\text{minimize} \quad g(\vec{y}) = \sum_{i=1}^{m} b_i y_i + d$$

$$\text{subject to} \quad \sum_{i=1}^{m} a_{ij} y_i \ge c_j, \quad 1 \le j \le n.$$

Thus, the coefficients in the dual objective function are the right-hand sides of the primal constraints; the m-tuple of coefficients in the jth dual constraint is the jth column of the coefficient matrix A; the right-hand sides of the dual constraints are the coefficients in the primal objective function.

Note that we can construct the primal from the dual as easily as the dual from the primal. As stated before, the two problems form a dual/primal pair—mathematically, they are equally important. Their *interpretations* may make one of them more important to us than the other.

For Problem (3.2), the coefficient matrix is

$$A = \begin{pmatrix} 1 & 1 & 1 \end{pmatrix}.$$

Also
$$\vec{c} = (\ -3 \quad 2 \quad 1\ ), \qquad \vec{b} = (1).$$
Thus, the corresponding dual problem is

$$\text{minimize} \quad y_1 \tag{3.6}$$
$$\text{subject to} \quad y_1 \geq -3$$
$$y_1 \geq 2$$
$$y_1 \geq 1.$$

The dual problem corresponding to Problem (3.5) is

$$\text{minimize} \quad 2y_1 - y_2 \tag{3.7}$$
$$\text{subject to} \quad y_1 - y_2 \geq 1$$
$$y_1 - 3y_2 \geq 0$$
$$y_1 - y_2 \geq 0$$
$$-y_1 + 2y_2 \geq -1.$$

This problem is feasible since $\vec{y} = (1, 0)$ is a feasible vector.

The solutions of the dual and primal problems are closely related. The following theorem reveals part of the relationship.

THEOREM 3.1. *If both the primal problem and its corresponding dual are feasible, then*
$$f(\vec{x}) \leq g(\vec{y}),$$
for any feasible vector $\vec{x}$ of the primal and any feasible vector $\vec{y}$ of the dual. Thus,
$$\max f(\vec{x}) \leq \min g(\vec{y}).$$

PROOF. Compute

$$
\begin{aligned}
f(\vec{x}) &= \sum_{j=1}^{n} c_j x_j + d \\
&\leq \sum_{j=1}^{n} \sum_{i=1}^{m} a_{ij} y_i x_j + d \\
&= \sum_{i=1}^{m} (\sum_{j=1}^{n} a_{ij} x_j) y_i + d \\
&\leq \sum_{i=1}^{m} b_i y_i + d \\
&= g(\vec{y}).
\end{aligned}
$$

The second inequality is obtained by maximizing over $\vec{x}$ on the left and then minimizing over $\vec{y}$ on the right.   □

From this, we obtain the following:

COROLLARY 3.2. *The following statements hold.*

(1) *If both the primal and its corresponding dual are feasible, then both are bounded.*
(2) *If the primal is unbounded, then the dual is infeasible.*
(3) *If the dual is unbounded, then the primal is infeasible.*

PROOF. For a fixed feasible vector $\vec{y}$ for the dual, $f(\vec{x})$ is bounded above by $g(\vec{y})$, and thus the primal problem is bounded. Similarly, the dual problem is bounded. The other two statements follow easily from the first.   □

As an example, we noted that both Problem (3.5) and its dual, Problem (3.7), are feasible. Thus, they are both bounded. For a second example, note that Problem (3.6) is easy to solve. It has only one unknown and the second constraint implies the other two. Hence, the problem calls for us to minimize y_1 subject only to the constraint $y_1 \geq 2$. Obviously, 2 is the minimum value of the objective function and it is attained for $y_1 = 2$. Now, Problem (3.6) is the dual corresponding to Problem (3.2). From Theorem 3.1 and its corollary, we conclude that Problem (3.2) is bounded and that its objective function is bounded above by 2 on the set of feasible vectors. We can go one step further. The vector $\vec{x} = (0,1,0)$ is a feasible vector for Problem (3.2) and the objective function has value 2 at this feasible vector. Since the objective function can be no larger than 2, we have solved Problem (3.2).

Exercises

(1) Consider the problem

$$\text{minimize} \quad -3x_1 + 2x_2 - x_3$$
$$\text{subject to} \quad x_1 + x_2 + x_3 = 3$$
$$x_1 - x_2 \geq 0.$$

Convert it into an equivalent primal problem.

(2) Consider the problem

$$\text{maximize} \quad x_1 - x_2 - x_3 + x_4$$
$$\text{subject to} \quad x_1 - x_3 \leq 0$$
$$x_2 - x_4 \leq 1.$$

Is it feasible? Is it bounded?

(3) Consider the dual problem

$$\text{minimize} \quad y_1 - 2y_2 + y_3$$

$$\text{subject to} \quad -y_1 - y_2 + y_3 \geq -1.$$

Is it feasible? Is it bounded? Write down the corresponding primal problem. Is the primal feasible? Is the primal bounded?

(4) Solve the problem

$$\text{maximize} \quad x_1 + x_2 + x_3$$

$$\text{subject to} \quad x_1 + 2x_2 + 3x_3 \leq 10.$$

(5) Consider the problem

$$\text{maximize} \quad -x_1 + 2x_2 - 3x_3 + 4x_4$$

$$\text{subject to} \quad x_3 - x_4 \leq 0$$

$$x_1 - 2x_3 \leq 1$$

$$2x_2 + x_4 \leq 3$$

$$-x_1 + 3x_2 \leq 5.$$

Is it feasible? Is it bounded?

(6) Consider the problem

$$\text{maximize} \quad x_1 - 3x_2$$

$$\text{subject to} \quad x_1 + x_2 \geq 1$$

$$x_1 - x_2 \leq 1$$

$$x_1 - x_2 \geq -1.$$

Is it feasible? Is it bounded?

3.2. Basic Forms and Pivots

Consider a primal LP problem:

$$\text{maximize} \quad f(\vec{x}) = \sum_{j=1}^{n} c_j x_j + d$$

$$\text{subject to} \quad \sum_{j=1}^{n} a_{ij} x_j \leq b_i, \quad 1 \leq i \leq m.$$

For each of the m constraints, define a *slack variable* by

$$x_{n+i} = b_i - \sum_{j=1}^{n} a_{ij} x_j.$$

Thus, the m slack variables depend on the unknowns $x_1, \ldots, x_n$, and each x_{n+i} measures the extent to which the left-hand side of constraint number

i is less than the right-hand side. Notice also that $\vec{x} = (x_1, \ldots, x_n)$ is a feasible vector if and only if

$$x_k \geq 0, \quad 1 \leq k \leq n + m.$$

Using these slack variables, we rewrite the problem in a way which appears strange at first but which turns out later to be convenient. It is

$$\text{maximize} \quad f(\vec{x}) = \sum_{j=1}^{n} c_j x_j + d \qquad (3.8)$$

$$\text{subject to} \quad \sum_{j=1}^{n} a_{ij} x_j - b_i = -x_{n+i}, \quad 1 \leq i \leq m.$$

This form of the primal problem is called the (*primal*) *basic form*. We emphasize that it is mathematically equivalent to the original primal form in the sense that solving one is the same as solving the other.

Let us write out a basic form for Problem (3.5). There are two constraints and so we have two slack variables, x_5 and x_6. The basic form is

$$\text{maximize} \quad x_1 - x_4 \qquad (3.9)$$

$$\text{subject to} \quad x_1 + x_2 + x_3 - x_4 - 2 = -x_5$$

$$-x_1 - 3x_2 - x_3 + 2x_4 + 1 = -x_6.$$

In the context of basic forms, the variables on the right-hand side of the constraints (that is, $x_{n+1}, \ldots, x_{n+m}$) are called *basic* (or, sometimes, *dependent*) variables while the other variables are called *nonbasic* (or *independent*) variables. The set of basic variables is called the *basis* corresponding to the basic form.

3.2.1. *Pivots*

With the use of a little simple algebra, we can compute a new basic form from an old one. To do so, choose, in Basic Form (3.8), any nonzero coefficient a_{kl}, where $1 \leq k \leq m$ and $1 \leq l \leq n$. Thus a_{kl} is the coefficient of x_l in the equation for $-x_{n+k}$. Since $a_{kl} \neq 0$, we can solve the equation to get x_l in terms of x_{n+k} and the remaining nonbasic variables (other than x_l). Then we substitute for x_l in the other constraints and in the objective function. The result is another basic form, differing from the first in that a basic and nonbasic variable have traded places. The operation we have just carried out is called a *pivot*, and we say that we *pivot on* the coefficient a_{kl}. It is important to realize that the new basic form is again mathematically equivalent to the old one, and thus to the original primal problem.

Let us pivot on the coefficient of x_3 in the equation for $-x_6$ in Basic Form (3.9). Solving for x_3 in that equation gives us

$$x_3 = -x_1 - 3x_2 + x_6 + 2x_4 + 1.$$

Substituting this expression for x_3 in the equation for $-x_5$ and combining terms give us

$$-2x_2 + x_6 + x_4 - 1 = -x_5.$$

The formula for the objective function in Basic Form (3.9) does not contain x_3 and so is not changed. The new basic form is

$$\text{maximize} \quad x_1 - x_4 \tag{3.10}$$

$$\text{subject to} \quad -2x_2 + x_6 + x_4 - 1 = -x_5$$

$$x_1 + 3x_2 - x_6 - 2x_4 - 1 = -x_3.$$

We carry out one more pivot, this time on the coefficient of x_1 in the equation for $-x_3$. The result is

$$\text{maximize} \quad -x_3 - 3x_2 + x_6 + x_4 + 1 \tag{3.11}$$

$$\text{subject to} \quad -2x_2 + x_6 + x_4 - 1 = -x_5$$

$$x_3 + 3x_2 - x_6 - 2x_4 - 1 = -x_1.$$

Thus the basis corresponding to Basic Form (3.9) is $\{x_5, x_6\}$, the basis corresponding to Basic Form (3.10) is $\{x_5, x_3\}$, and the basis corresponding to Basic Form (3.11) is $\{x_5, x_1\}$. It is clear that there are, in general, many basic forms for each primal problem. Any finite sequence of pivots transforms one basic form into another. In the general case where there are n unknowns in the primal problem and m constraints, the total number of possible bases corresponding to basic forms is the binomial coefficient

$$\binom{m+n}{m} = \frac{(m+n)!}{n!m!}.$$

For example, if $n = 4$ and $m = 2$, this number is 15. The actual number of basic forms might be less than this quantity.

Given a basic form, the basic variables and the objective function are expressed in terms of the nonbasic variables. If arbitrary values are assigned to the nonbasics, then the values of the basic variables are determined, and so we obtain an $(n + m)$-tuple $(x_1, \ldots, x_{n+m})$. In case each $x_k \geq 0$, we say that this $(n + m)$-tuple is *feasible*. If $(x_1, \ldots, x_{n+m})$ is feasible, then the vector $(x_1, \ldots, x_n)$ is, as already noted, a feasible vector.

The most important case of this computation occurs when the value zero is assigned to each nonbasic. We have

DEFINITION 3.3. Given a basic form for a primal problem with n unknowns and m constraints, the corresponding *basic solution* is the $(m+n)$-tuple obtained by setting the n nonbasic variables to zero and then solving for the m basic variables from the constraints.

For example, in Basic Form (3.9), the basic solution is $(0, 0, 0, 0, 2, -1)$. Note that the value of a basic variable in a basic solution is just the negative of the constant term in its constraint equation. In Basic Forms (3.10) and (3.11), the basic solutions are $(0, 0, 1, 0, 1, 0)$ and $(1, 0, 0, 0, 1, 0)$, respectively. Also note that the value of the objective function at the basic solution is the constant term in its equation.

DEFINITION 3.4. A basic form is *feasible* if its basic solution is feasible. A feasible basic solution is *optimal* if the value of the objective function at the basic solution is a maximum. In this case, the basic form is also called *optimal*.

We know that a vector $(x_1, \ldots, x_n)$ is a feasible vector for the primal problem if and only if $(x_1, \ldots, x_{n+m})$ is feasible. It will be proved later that if a problem has an optimal feasible vector then it has an optimal feasible basic solution.

We see, for example, that Basic Form (3.9) is infeasible while Basic Forms (3.10) and (3.11) are both feasible.

In the next section, we shall describe the *simplex algorithm*, which is a numerical method for solving primal problems. The idea of it is to start with a feasible basic form and then to pivot in such a way that the new basic form is still feasible and gives a larger value of the objective function. This process is repeated until the maximum is reached. The algorithm includes a simple method for deciding which coefficient to pivot on next.

In order to recognize feasible and optimal basic forms, we need

THEOREM 3.3. *The following hold:*

(1) *A basic form is feasible if and only if the constant term in each constraint is nonpositive.*

(2) *Suppose a basic form is feasible. If each coefficient (of a nonbasic variable) in the equation for the objective function is nonpositive, then the basic form is optimal.*

PROOF.   (1) This is clear since the value of a basic variable in the basic solution is the negative of the constant term in its equation.

(2) We can write the objective function as

$$f(\vec{x}) = \sum_{k=1}^{m+n} c_k x_k + d,$$

where $c_k = 0$ if x_k is basic (since basic variables do not appear in the equation for the objective function). Our hypothesis is that $c_k \leq 0$ for each k. Now, if $\vec{z}$ is the basic solution, we have

$$f(\vec{z}) = d.$$

Then if $\vec{x}$ is *any* feasible vector, we have, since each $x_k \geq 0$,

$$f(\vec{x}) = \sum_{k=1}^{m+n} c_k x_k + d \leq d = f(\vec{z}).$$

Thus, $\max f(\vec{x}) = f(\vec{z})$.

□

We can illustrate the second part of the theorem by doing a pivot on Basic Form (3.11). In fact, if we pivot on the coefficient of x_4 in the equation for $-x_5$, we get

$$\text{maximize} \quad -x_3 - x_2 - x_5 + 2 \tag{3.12}$$

$$\text{subject to} \quad -2x_2 + x_6 + x_5 - 1 = -x_4$$

$$x_3 - x_2 + x_6 + 2x_5 - 3 = -x_1.$$

According to the theorem, this basic form is both feasible and optimal and thus we have solved Problem (3.5). To summarize, the solution is

$$\max f(\vec{x}) = 2, \quad x_1 = 3, \quad x_2 = x_3 = 0, \quad x_4 = 1.$$

3.2.2. *Dual Basic Forms*

Consider a dual problem

$$\text{minimize} \quad \sum_{i=1}^{m} b_i y_i + d$$

$$\text{subject to} \quad \sum_{i=1}^{m} a_{ij} y_i \geq c_j, \quad 1 \leq j \leq n.$$

For each of the n constraints, define a *surplus variable* by

$$y_{m+j} = \sum_{i=1}^{m} a_{ij} y_i - c_j.$$

Then y_{m+j} measures the difference between the left and right sides of the jth constraint. We see that $(y_1, \ldots, y_m)$ is a feasible vector for the dual problem if and only if $y_k \geq 0$ for $1 \leq k \leq m + n$. We rewrite the dual problem as

$$\text{minimize} \quad \sum_{i=1}^{m} b_i y_i + d$$

$$\text{subject to} \quad y_{m+j} = \sum_{i=1}^{m} a_{ij} y_i - c_j, \quad 1 \leq j \leq n.$$

This is a *dual basic form*.

The variables on the left-hand side of the equality signs are *basic* and the others are *nonbasic*. Just as in the primal case, we can *pivot* so as to interchange a basic and a nonbasic variable and produce a new dual basic form.

Let us write down the dual basic form for Problem (3.7). It is

$$\text{minimize} \quad 2y_1 - y_2 \tag{3.13}$$

$$\text{subject to} \quad y_3 = y_1 - y_2 - 1$$

$$y_4 = y_1 - 3y_2$$

$$y_5 = y_1 - y_2$$

$$y_6 = -y_1 + 2y_2 + 1.$$

If we pivot on the coefficient of y_1 in the equation for y_3, we get

$$\text{minimize} \quad 2y_3 + y_2 + 2 \tag{3.14}$$

$$\text{subject to} \quad y_1 = y_3 + y_2 + 1$$

$$y_4 = y_3 - 2y_2 + 1$$

$$y_5 = y_3 + 1$$

$$y_6 = -y_3 + y_2.$$

Given a dual basic form, the definition of *feasible* $(n + m)$-tuple is the same as in the case of primal basic forms. Also, the *basic solution* corresponding to a dual basic form is obtained by setting the nonbasic variables equal to zero, and then solving for the basic variables. Clearly, the value of a basic variable in a basic solution is simply the constant term in its equation. We define a basic solution $(y_1, \ldots, y_{m+n})$ (and its dual basic form) to be *feasible* if $y_k \geq 0$ for $1 \leq k \leq m+n$. The value of the objective function at the basic solution is clearly equal to the constant term in its equation. A feasible basic solution (and its dual basic form) is *optimal* if the value of the objective function is a minimum.

In Dual Basic Form (3.13), the basic solution is $(0, 0, -1, 0, 0, 1)$; it is infeasible. In Dual Basic Form (3.14), the basic solution is $(1, 0, 0, 1, 1, 0)$; it is feasible.

A theorem analogous to Theorem 3.3 holds for dual basic forms. The proof is omitted since it is very similar to the one for Theorem 3.3.

THEOREM 3.4. *The following are true:*

(1) *A dual basic form is feasible if and only the constant term in each constraint equation is nonnegative.*

(2) *Suppose a dual basic form is feasible. If each coefficient (of a non-basic variable) in the equation for the objective function is nonnegative then the dual basic form is optimal.*

This theorem shows that Dual Basic Form (3.14) is both feasible and optimal. Thus the solution to Problem (3.7) is

$$\min g(\vec{y}) = 2, y_1 = 1, y_2 = 0.$$

Exercises

(1) Consider the problem

$$\text{maximize} \quad 2x_1 + 3x_2$$

$$\text{subject to} \quad x_1 + x_2 \leq 10$$

$$x_1 - x_2 \leq 2.$$

Write down *all* the basic forms.

(2) Consider the primal problem

$$\text{maximize} \quad x_1 - 2x_2$$

$$\text{subject to} \quad x_1 \leq 2$$

$$-x_1 + x_2 \leq 3.$$

Write down *all* the basic forms. Which are feasible? Which is optimal?

(3) Consider the dual problem

$$\text{minimize} \quad y_1 + y_2$$

$$\text{subject to} \quad 2y_1 + 3y_2 \geq 5$$

$$-y_1 \geq -6.$$

Write down *all* the dual basic forms. Which are feasible? Which is optimal?

(4) Consider the problem

$$\text{minimize} \quad -y_1 + 2y_2 + y_3$$

$$\text{subject to} \quad y_1 - y_2 \geq 0$$

$$-y_1 + y_3 \geq 1$$

$$-y_2 + 2y_3 \geq -1.$$

Carry out *one* pivot to reach an optimal basic form.

(5) Solve the problem

$$\text{maximize} \quad x_1 - x_2 + 2x_3$$
$$\text{subject to} \quad x_1 + x_2 + x_3 \leq 3.$$

3.3. The Simplex Algorithm

In order to discuss the simplex algorithm, it is convenient to introduce a more compact way of writing a basic form. It is called a *tableau* and is discussed next.

3.3.1. *Tableaus*

Let us consider an example, namely Basic Form (3.11). The *tableau* corresponding to it is

$$
\begin{array}{cccc|c|c}
x_3 & x_2 & x_6 & x_4 & -1 & \\
\hline
0 & -2 & 1 & 1 & 1 & = -x_5 \\
1 & 3 & -1 & -2 & 1 & = -x_1 \\
\hline
-1 & -3 & 1 & 1 & -1 & = \quad f
\end{array}
\tag{3.15}
$$

This tableau is essentially a 3×5 matrix with *labels* on the rows and columns. In the general case, where the primal problem has n unknowns and m constraints, the size would be $(m+1) \times (n+1)$. In our example, the first four columns are labeled by the nonbasic variables, x_3, x_2, x_6, x_4. The fifth column is labeled -1. This label will be explained shortly. The first two rows are labeled on the right by the basic variables "$= -x_5, = -x_1$." The last row is labeled by the objective function "$= f$." Each row is to be read as an equation—the numbers in the row are multiplied by the labels at the tops of the corresponding columns, and these products are added. Thus, the first row represents the equation

$$(0)(x_3) + (-2)(x_2) + (1)(x_6) + (1)(x_4) + (1)(-1) = -x_5.$$

This simplifies to

$$-2x_2 + x_6 + x_4 - 1 = -x_5,$$

which is the first constraint in Basic Form (3.11). The second row gives the second constraint and the bottom row gives the objective function.

The basic solution is easily read off the tableau. In fact, the values of the basic variables are simply the numbers in the right-hand column (excluding the bottom row); the value of the objective function at the basic solution is the negative of the number in the lower right-hand corner. Theorem 3.3 gives conditions on the coefficients in a basic form to ensure feasibility and optimality. This result can be easily translated into the context of tableaus.

THEOREM 3.5. *Consider a tableau of size* $(m+1) \times (n+1)$. *Then we have:*

(1) *The tableau is feasible if and only if each entry in the right-hand column (not counting the bottom one) is nonnegative.*

(2) *Suppose the tableau is feasible. If each entry in the bottom row (not counting the right-hand one) is nonpositive, then the tableau is optimal.*

The result of pivoting on the coefficient of x_4 in the equation for $-x_5$ in Basic Form (3.11) is given in Basic Form (3.12). The corresponding tableau is

$$
\begin{array}{cccc|c|cc}
x_3 & x_2 & x_6 & x_5 & -1 & & \\
\hline
0 & -2 & 1 & 1 & 1 & = & -x_4 \\
1 & -1 & 1 & 2 & 3 & = & -x_1 \\
\hline
-1 & -1 & 0 & -1 & -2 & = & f
\end{array}
\qquad (3.16)
$$

The effect of a pivot on the labels is simply to interchange a basic and a nonbasic variable. In the example, basic variable x_5 and nonbasic variable x_4 trade places. The next theorem shows how to compute the numbers in the new tableau from those in the old one.

THEOREM 3.6. *Let T be a tableau of size $(m+1) \times (n+1)$. Suppose a pivot is carried out on entry t_{kl}. In other words, the pivot is on the coefficient of the nonbasic variable x_p (labeling column l) in the constraint equation for the basic variable x_q (labeling row k). Of course, $1 \le k \le m$, $1 \le l \le n$, and $t_{kl} \ne 0$. Let T' be the tableau resulting from the pivot. Then, if $1 \le i \le m+1$ and $1 \le j \le n+1$, we have:*

(1) $t'_{kl} = 1/t_{kl}$.

(2) *If $j \ne l$, then $t'_{kj} = t_{kj}/t_{kl}$.*

(3) *If $i \ne k$, then $t'_{il} = -t_{il}/t_{kl}$.*

(4) *If $i \ne k$, and $j \ne l$, then*

$$
t'_{ij} = \frac{t_{ij}t_{kl} - t_{il}t_{kj}}{t_{kl}}.
$$

PROOF. We prove only (4), since the other cases are similar but easier. To do this, let α be the label on column j. Thus α is either one of the nonbasic variables other than x_p, or the constant -1. Also, let β be the label on row i. Thus β is either the negative of a basic variable other than x_q, or the objective function. Now, solving for x_p from the constraint equation for $-x_q$, we have

$$
x_p = -\left(\frac{t_{kj}}{t_{kl}}\right)\alpha + \Gamma,
\qquad (3.17)
$$

where Γ is a sum of terms not involving α. Then, from row i of T, we have

$$\beta = t_{il}x_p + t_{ij}\alpha + \Delta,$$

where Δ is a sum of terms not involving x_p or α. Substituting from (3.17), we have

$$\beta = t_{il}[-\left(\frac{t_{kj}}{t_{kl}}\right)\alpha + \Gamma] + t_{ij}\alpha + \Delta.$$

Thus,

$$\beta = \left(\frac{t_{ij}t_{kl} - t_{il}t_{kj}}{t_{kl}}\right)\alpha + \Theta,$$

where Θ consists of terms not involving α. Since t'_{ij} is the coefficient of α in the equation for β, the proof is complete. $\square$

Let us check the theorem on the example already discussed. The pivot carried out on Tableau (3.15) to produce Tableau (3.16) is on entry $t_{14} = 1$. From parts (1) and (2) of the theorem, we see that row 1 of Tableau (3.16) should be identical to row 1 of Tableau (3.15). Indeed, it is. The entries in column 4 of Tableau (3.16), other than t_{14}, should be the negatives of the entries in Tableau (3.15). This comes from part (3) of the theorem. Again, this is true. To check part (4) of the theorem, consider entry t'_{32} [of Tableau (3.16)]. It should be

$$t'_{32} = \frac{t_{32}t_{14} - t_{12}t_{34}}{t_{14}} = \frac{(-3)(1) - (-2)(1)}{1} = -1.$$

This is correct.

We summarize the rules for pivoting in a tableau in the following algorithm.

ALGORITHM 3.1 (THE PIVOTING ALGORITHM). *We are given a tableau* T *of size* $(m+1) \times (n+1)$, *and an entry* $t_{kl} \neq 0$. *We are to compute a new tableau* T' *of the same size as* T.

(1) *Interchange the labels on column* l *and row* k. *The other labels on* T' *are the same as on* T.
(2) $t'_{kl} = 1/t_{kl}$.
(3) *If* $j \neq l$, *then* $t'_{kj} = t_{kj}/t_{kl}$.
(4) *If* $i \neq k$, *then* $t'_{il} = -t_{il}/t_{kl}$.
(5) *If* $i \neq k$ *and* $j \neq l$, *then*

$$t'_{ij} = \frac{t_{ij}t_{kl} - t_{il}t_{kj}}{t_{kl}}.$$

3.3.2. *The Simplex Algorithm*

Before stating the simplex algorithm, we need one more theorem about tableaus.

THEOREM 3.7. *Consider a feasible tableau for a primal problem with n unknowns and m constraints. Suppose that one of the first n entries in the bottom row of the tableau is positive, but that all the other entries in the column containing that entry are nonpositive. Then the problem is unbounded.*

PROOF. Suppose that it is column j which has a positive entry at the bottom and which has all nonpositive entries otherwise. Then column j is labeled by some nonbasic variable, say x_q. Given any positive number x, define an $(m+n)$-tuple as follows: To x_q, assign the value x; to a nonbasic x_l, $l \neq q$, assign the value 0; then solve for the basics from their constraint equations. Then each basic variable in this $(m+n)$-tuple is nonnegative because of the condition on the entries in column j. Since the coefficient of x_q in the objective function is positive, the objective function has limit $+\infty$ as $x \to \infty$. Thus, the objective function is unbounded. ☐

For example, consider the problem

$$\text{maximize} \quad x_1 \tag{3.18}$$

$$\text{subject to} \quad x_1 - x_2 \leq 1.$$

The initial tableau is

x_1	x_2	-1		
1	-1	1	$=$	$-x_3$
1	0	0	$=$	f

This is feasible and nonoptimal. The theorem does not indicate that it is unbounded. We now pivot on the coefficient of x_1 in the constraint equation for $-x_3$. The new tableau is

x_3	x_2	-1		
1	-1	1	$=$	$-x_1$
-1	1	-1	$=$	f

This is feasible. The theorem tells us that the problem is unbounded.

ALGORITHM 3.2 (THE SIMPLEX ALGORITHM). *We are given a feasible tableau T of size $(m+1) \times (n+1)$.*

(1) *If*

$$t_{m+1,j} \leq 0 \quad \text{for } 1 \leq j \leq n,$$

then STOP (the tableau is optimal).

(2) *Choose any l with $1 \leq l \leq n$ such that $t_{m+1,l} > 0$.(The nonbasic variable labeling column l is called the* entering *variable.)*

(3) *If*

$$t_{il} \leq 0 \quad for \; 1 \leq i \leq m,$$

STOP (the problem is unbounded).

(4) *Choose any k with $1 \leq k \leq m$ such that $t_{kl} > 0$ and*

$$\frac{t_{k,n+1}}{t_{kl}} = \min \left\{ \frac{t_{i,n+1}}{t_{il}} : 1 \leq i \leq m \; and \; t_{il} > 0 \right\}.$$

(The basic variable labeling row k is called the leaving *variable.)*

(5) *Pivot on t_{kl} and replace T by the tableau resulting from this pivot. (Entry t_{kl} is called the* pivot entry.)

(6) *Go to Step (1).*

In the next theorem, we verify that the algorithm produces a sequence of feasible tableaus such that the values of the objective function are non-decreasing.

THEOREM 3.8. *The following hold for the simplex algorithm:*

(1) *After each pivot [Step (5)], the new tableau is feasible.*

(2) *After each pivot, the value of the objective function for the new tableau is greater than or equal to that for the previous tableau.*

PROOF.   (1) We are pivoting on entry t_{kl} of T to produce T'. We must verify that

$$t'_{i,n+1} \geq 0 \quad for \; 1 \leq i \leq m.$$

This is clearly true for $i = k$, so assume $i \neq k$. Then

$$t'_{i,n+1} = \frac{t_{i,n+1}t_{kl} - t_{il}t_{k,n+1}}{t_{kl}}.$$

Now, we know that $t_{i,n+1}, t_{kl}, t_{k,n+1}$ are all nonnegative. There are two possibilities. If $t_{il} \leq 0$ then $t'_{i,n+1} \geq 0$. On the other hand, if $t_{il} > 0$ then, by the choice of k,

$$\frac{t_{k,n+1}}{t_{kl}} \leq \frac{t_{i,n+1}}{t_{il}}.$$

Multiplying both sides by $t_{kl}t_{il}$, we get

$$t_{il}t_{k,n+1} \leq t_{i,n+1}t_{kl}.$$

This implies that

$$t'_{i,n+1} \geq 0.$$

(2) The value of the objective function for T is $-t_{m+1,n+1}$; for T', it is $-t'_{m+1,n+1}$. Now

$$t'_{m+1,n+1} = \frac{t_{m+1,n+1}t_{kl} - t_{m+1,l}t_{k,n+1}}{t_{kl}} \leq \frac{t_{m+1,n+1}t_{kl}}{t_{kl}} = t_{m+1,n+1}.$$

Taking negatives on both sides completes the proof. □

There are two things about the algorithm which will be cleared up in the next section. The first is the problem of producing an initial feasible tableau (without which the algorithm cannot be started). The second is the problem of whether the algorithm eventually stops with the right answer. It *is* clear that, if it stops, either the problem is unbounded or the solution has been reached.

For an example, consider the primal problem,

$$\text{maximize} \quad x_1 - x_2 + x_3 - 2x_4 \tag{3.19}$$

$$\text{subject to} \quad x_1 + x_2 + x_3 + x_4 \leq 4$$

$$-x_1 + x_3 \leq 0$$

$$x_2 - x_4 \leq 1.$$

The initial tableau is then

x_1	x_2	x_3	x_4	-1		
1	1	1	1	4	$=$	$-x_5$
-1	0	*1	0	0	$=$	$-x_6$
0	1	0	-1	1	$=$	$-x_7$
1	-1	1	-2	0	$=$	f

This tableau is feasible but not optimal; unboundedness is not indicated. The algorithm allows two choices of entering variable: x_1 or x_3. We choose x_3. In the column labeled by x_3, there are two positive entries above the bottom row. The minimum ratio is zero and is achieved by the entry in the second row. Thus, x_3 will enter the basis and x_6 will leave it. The pivot entry is marked with an "$*$." We pivot to get the tableau

x_1	x_2	x_6	x_4	-1		
*2	1	-1	1	4	$=$	$-x_5$
-1	0	1	0	0	$=$	$-x_3$
0	1	0	-1	1	$=$	$-x_7$
2	-1	-1	-2	0	$=$	f

This tableau is feasible (as it must be) but not optimal; unboundedness is not indicated. There is only one possible pivot entry: x_1 must enter the basis and x_5 must leave it. The pivot entry is again marked. We pivot to get

x_5	x_2	x_6	x_4	-1		
1/2	1/2	$-1/2$	1/2	2	=	$-x_1$
1/2	1/2	1/2	1/2	2	=	$-x_3$
0	1	0	-1	1	=	$-x_7$
-1	-2	0	-3	-4	=	f

This tableau is optimal. We get the solution

$$\max f(\vec{x}) = 4, \quad x_1 = 2, \quad x_2 = 0, \quad x_3 = 2, \quad x_4 = 0.$$

In fact, it has more than one solution. To see this, pivot in the last tableau so as to interchange x_6 and x_3.

Exercises

(1) Solve

$$\text{maximize} \quad x_1$$
$$\text{subject to} \quad 2x_1 - x_2 \leq 2$$
$$x_2 + x_3 \leq 2.$$

(2) Solve

$$\text{maximize} \quad x_1 + 2x_2 + 3x_3$$
$$\text{subject to} \quad x_1 + x_2 + x_3 \leq 3$$
$$-x_1 + x_2 \leq 0.$$

(3) Solve

$$\text{maximize} \quad x_1 + x_2 + x_3$$
$$\text{subject to} \quad x_1 + x_3 \leq 3$$
$$x_2 - x_4 \leq 0.$$

(4) Solve

$$\text{maximize} \quad x_1 + x_2 - x_3 - 2x_4$$
$$\text{subject to} \quad x_1 + 2x_2 + 2x_3 + x_4 \leq 10$$
$$x_1 - x_3 \leq 0$$
$$x_2 \leq 2.$$

(5) Verify that Problem (3.19) can be solved in one pivot.

(6) Find a condition on an optimal tableau which guarantees that there are at least *two* solutions.

3.4. Avoiding Cycles and Achieving Feasibility

As the simplex algorithm was stated in the previous section, it is possible for it to go into an infinite loop and never deliver a solution to the problem. There are several methods for avoiding this difficulty. We will present one of them. After that, we will discuss the question of how to find an initial feasible tableau so that the algorithm can be started.

3.4.1. *Degeneracy and Cycles*

First, we have the following:

DEFINITION 3.5. Let T be an $(m+1) \times (n+1)$ tableau. The pivot on entry t_{kl} is said to be *degenerate* if $t_{k,n+1} = 0$.

The first pivot we carried out in solving Problem (3.19) was degenerate. The important fact about degenerate pivots is given in the following theorem. The proof should be obvious.

THEOREM 3.9. *If the tableau T' is obtained from T by a degenerate pivot then*

$$t'_{i,n+1} = t_{i,n+1},$$

for all i with $1 \le i \le m+1$.

In other words, the last column of T' is identical to the last column of T.

COROLLARY 3.10. *If T' is obtained from T by a degenerate pivot, then the basic solution corresponding to T' is equal to the basic solution corresponding to T. Thus, the objective function has the same value for the two basic solutions.*

PROOF. Since the pivot is degenerate, the value of the leaving variable in T is zero. This is also its value in T' since it is nonbasic. The value of the entering variable remains zero; the values of the others clearly do not change. $\square$

Thus, doing a degenerate pivot while carrying out the simplex algorithm causes no improvement in the objective function. This is usually harmless [as Problem (3.19) shows]. Degenerate pivots are eventually followed by nondegenerate ones, and the progression toward a solution continues. There is, however, a rare event which must be allowed for in the theory. It is called *cycling* and can put the algorithm into an infinite loop. A *cycle* consists of a sequence of feasible tableaus $T^0, T^1, \ldots, T^k$, such that each is obtained from the previous one by a degenerate pivot chosen according to the algorithm, and such that T^k has the same basis as T^0. Thus, T^k can differ from T^0 only in the order of its rows and columns. Now, if the pivot in T^k is chosen just as the one in T^0 was, the result can be that the computation repeats the list of tableaus ad infinitum.

As we have said, cycling is rare. There are examples involving reasonably small problems (an example is given in [Str89]), but only a few have have ever arisen which were not intentionally made up to illustrate the phenomenon. For the theory, however, cycling is a serious problem because it blocks a proof that the algorithm stops. Fortunately, there is a simple way to avoid the problem. It first appeared in [Bla77] and is an amendment to the simplex algorithm. It is the following:

BLAND'S ANTICYCLING RULE

Modify the simplex algorithm so that, whenever there is a choice of entering or leaving variable, we always choose the one with smallest subscript.

If we had applied Bland's rule in Problem (3.19), the first pivot would have interchanged x_1 and x_5 instead of x_3 and x_6. Thus, the degenerate pivot would have been eliminated. In general, Bland's rule does not eliminate all degenerate pivots; it does, however, prevent cycles.

THEOREM 3.11. *If the simplex algorithm is modified by Bland's rule, then no cycle occurs.*

PROOF. The strategy of the proof is to assume that there is a cycle $T^0, \ldots, T^k$, with pivots chosen according to Bland's rule, and derive a contradiction. Define t to be the largest member of the set

$$C = \{i : x_i \text{ enters the basis during the cycle}\}.$$

Thus C consists of all i such that x_i is basic for some tableaus in the cycle, and nonbasic for others. Since the bases for T^0 and T^k are identical, there is a tableau T^u in the cycle such that x_t is basic in T^u and nonbasic in T^{u+1}. Also, there is a tableau T^v such that x_t is nonbasic in T^v and basic in T^{v+1}. Let x_s be the variable which is nonbasic in T^u and basic in T^{u+1}. Thus x_s and x_t trade places in the pivot in T^u which produces T^{u+1}. Note that $s < t$ since s is in C.

Now, the objective function equation in T^u has the form

$$f(\vec{x}) = d + \sum_{k=1}^{n+m} c_k x_k, \tag{3.20}$$

where $c_k = 0$ whenever x_k is basic in T^u. Similarly, the objective function equation in T^v has the form

$$f(\vec{x}) = d + \sum_{k=1}^{n+m} c_k^* x_k, \tag{3.21}$$

where $c_k^* = 0$ whenever x_k is basic in T^v. The constant terms in (3.20) and (3.21) are the same because the value of the objective function does not change after a degenerate pivot.

For any $(n + m)$-tuple $\bar{x}$ satisfying the constraints, the two formulas (3.20) and (3.21) must give the same value. That is, subtracting,

$$\sum_{k=1}^{n+m}(c_k - c_k^*)x_k = 0. \tag{3.22}$$

In particular, let us form $\bar{x}$ by setting $x_s = 1$, setting all the other variables nonbasic in T^u equal to 0, and solving for the variables basic in T^u from their constraint equations. We see that if x_i is basic in T^u, then the expression for x_i has the form

$$x_i = b_i - a_i.$$

In this equation, a_i is the coefficient of x_s in the constraint equation for $-x_i$, and b_i is the constant term in that equation. From (3.22), we get

$$c_s - c_s^* + \sum(-c_i^*)(b_i - a_i) = 0.$$

The sum here is over all variables basic in T^u, and we have used the fact that $c_i = 0$ when x_i is basic in T^u.

Now, $c_s^* \leq 0$ because, otherwise, x_s would be a nonbasic variable in T^v eligible to enter the basis. However, x_t actually enters the basis. Since $s < t$, Bland's rule is violated. Also, $c_s > 0$ since x_s enters the basis in the pivot on T^u. We conclude that

$$c_s - c_s^* > 0,$$

and so

$$\sum c_i^*(b_i - a_i) > 0.$$

We choose any r so that x_r is basic in T^u and

$$c_r^*(b_r - a_r) > 0.$$

It follows that $c_r^* \neq 0$ and so x_r is nonbasic in T^v. Thus $b_r = 0$. The last equation becomes

$$c_r^* a_r < 0.$$

Now, $r \neq t$ because $a_t > 0$ (it is the pivot entry in T^u) and $c_t^* > 0$ (x_t is the entering variable in T^v). Since x_r is basic in T^u and nonbasic in T^v, $r \in C$ and so $r < t$.

Finally, if $a_r > 0$, x_r is eligible to leave the basis in the pivot on T^u. Since $r < t$, the choice of x_t violates Bland's rule. Thus $a_r < 0$ and so $c_r^* > 0$. But this means that x_r is eligible to enter the basis in the pivot on T^v (instead of x_t). Again, Bland's rule is violated. $\square$

To state this theorem differently, if we modify the simplex algorithm by using Bland's rule, then no basis ever appears more than once. Since there are only finitely many bases, the algorithm must terminate after finitely many iterations.

There is still the problem of finding an initial feasible tableau.

3.4.2. *The Initial Feasible Tableau*

Let us consider the problem

$$\text{maximize} \quad 2x_1 - x_2 + 2x_3 \qquad\qquad (3.23)$$

$$\text{subject to} \quad x_1 + x_2 + x_3 \le 6$$

$$-x_1 + x_2 \le -1$$

$$-x_2 + x_3 \le -1.$$

Introducing slack variables gives us the basic form

$$\text{maximize} \quad 2x_1 - x_2 + 2x_3$$

$$\text{subject to} \quad x_1 + x_2 + x_3 - 6 = -x_4$$

$$-x_1 + x_2 + 1 = -x_5$$

$$-x_2 + x_3 + 1 = -x_6.$$

This basic form is infeasible, although the problem itself is feasible [$(2, 1, 0)$ is a feasible vector]. We introduce an *auxiliary* variable u, and use it to construct the *auxiliary problem*. It is

$$\text{maximize} \quad -u$$

$$\text{subject to} \quad x_1 + x_2 + x_3 - u - 6 = -x_4$$

$$-x_1 + x_2 - u + 1 = -x_5$$

$$-x_2 + x_3 - u + 1 = -x_6.$$

Notice that the auxiliary objective function has nothing whatsoever to do with the objective function for Problem (3.23). To understand why we are interested in the auxiliary problem, suppose that we can find a feasible tableau for it in which the auxiliary variable is nonbasic. The tableau obtained by erasing the column labeled by u is then a feasible tableau for Problem (3.23) (except that the original objective function must replace the auxiliary one). The reason for this is that $(x_1, \ldots, x_6)$ satisfies the original constraints if and only if $(x_1, \ldots, x_6, 0)$ satisfies the auxiliary constraints. (Here, the last entry in the second vector is u.)

To continue with the example, we write the tableau

x_1	x_2	x_3	u	-1		
1	1	1	-1	6	$=$	$-x_4$
-1	1	0	-1	-1	$=$	$-x_5$
0	-1	1	-1	-1	$=$	$-x_6$
0	0	0	-1	0	$=$	f

This tableau is infeasible, but it is easy to pivot it into a feasible one. In this example, the entry in column 4 and row 2 (or row 3) works. Carrying out the first of these pivots gives

x_1	x_2	x_3	x_5	-1		
2	0	1	-1	7	$=$	$-x_4$
1	-1	0	-1	1	$=$	$-u$
1	-2	1	-1	0	$=$	$-x_6$
1	-1	0	-1	1	$=$	f

This tableau is feasible. After two pivots chosen according to the simplex algorithm, we get

x_6	u	x_3	x_5	-1		
2	-4	3	1	3	$=$	$-x_4$
-1	1	-1	0	1	$=$	$-x_2$
-1	2	-1	-1	2	$=$	$-x_1$
0	-1	0	0	0	$=$	f

This is optimal and the auxiliary variable is nonbasic. In order to write down an initial feasible tableau for the original Problem (3.23), it is only necessary to erase the column labeled by u and to replace the bottom row by the coefficients of the original objective function (written in terms of the nonbasics x_6, x_3, and x_5). Since the objective function is given in terms of x_1, x_2, and x_3, we only have to replace x_1 and x_2 by the formulas for them given by the constraint equations. Thus

$$2x_1 - x_2 + 2x_3 = 2(x_6 + x_3 + x_5 + 2) - (x_6 + x_3 + 1) + 2x_3$$
$$= x_6 + 3x_3 + 2x_5 + 3.$$

The initial feasible tableau for Problem (3.23) is then

x_6	x_3	x_5	-1		
2	3	1	3	$=$	$-x_4$
-1	-1	0	1	$=$	$-x_2$
-1	-1	-1	2	$=$	$-x_1$
1	3	2	-3	$=$	f

The problem can then be solved in three pivots if we use Bland's rule, or in one if we do not.

We formalize this method as follows.

ALGORITHM 3.3 (THE FEASIBILITY ALGORITHM). *We are given an infeasible primal basic form. (This is called the* original *problem.)*

(1) *Introduce an auxiliary variable u.*

(2) *Define the auxiliary objective function to be −u.*

(3) *For each constraint in the original problem, form an auxiliary constraint by adding −u to the left-hand side.*

(4) *Define the auxiliary problem to be that of maximizing the auxiliary objective function subject to the auxiliary constraints.*

(5) *Set up the tableau for the auxiliary problem.*

(6) *Choose a row (not the bottom one) so that the right-hand entry is smallest (the most negative); pivot on the entry in this row and in the column labeled by u. (The result is a feasible tableau.)*

(7) *Apply the simplex algorithm to solve the auxiliary problem.*

(8) *If the maximum value of the auxiliary objective function is negative, STOP (the original problem is infeasible).*

(9) *Carry out a pivot so that the auxiliary variable is nonbasic (if necessary).*

(10) *Set up a feasible tableau for the original problem by erasing the column labeled by the auxiliary variable, and then replacing the auxiliary objective function by the original one (written in terms of the nonbasics).*

The reader is asked to verify as exercises the assertions made in Steps (6) and (8). Note that the auxiliary problem always has a solution since its objective function is bounded above by zero.

The only other point which seems to require comment is Step (9). It is conceivable that, after solving the auxiliary problem, we get a maximum of zero in a tableau in which the auxiliary variable is basic. Obviously, its value would be zero. It is, however, easy to see that a (degenerate) pivot makes it nonbasic.

The following theorem summarizes the results of the previous sections.

THEOREM 3.12. *If a primal linear programming problem is both feasible and bounded, then it has a solution. Moreover, a solution can be computed by using the feasibility algorithm (if necessary) together with the simplex algorithm, modified by Bland's rule.*

Exercises

(1) Solve

$$\text{maximize} \quad x_1 + x_3$$

$$\text{subject to} \quad x_1 + x_2 + x_3 \le 10$$

$$x_1 - x_2 - x_3 \le -1$$

$$x_1 \ge 2.$$

(2) Solve

$$\text{maximize} \quad x_1 + x_2 - x_3 - x_4$$

$$\text{subject to} \quad x_1 + x_2 + x_3 + x_4 \le 8$$

$$x_1 - x_2 - x_3 - x_4 \le -1$$

$$x_3 \le 4.$$

(3) Solve

$$\text{maximize} \quad 3x_1 - 2x_2 - x_3 + 5x_4$$

$$\text{subject to} \quad x_1 + 2x_2 + x_3 + 2x_4 \le 10$$

$$x_1 \le 1$$

$$x_4 \le 1$$

$$x_2 - x_3 \le -1$$

$$x_1 - x_4 \le -1.$$

(4) Prove that if the pivot is chosen as in Step (6) of the feasibility algorithm, then the result is a feasible tableau.

(5) Prove the assertion made in Step (8) of the feasibility algorithm.

3.5. Duality

In this section, we describe the dual versions of the results already obtained for primal problems. It will then be easy to finish the job begun earlier (in Theorem 3.1 and its corollary) of revealing the connection between primal and dual problems.

3.5.1. *The Dual Simplex Algorithm*

For an example, let us consider Dual Basic Form (3.13). The *dual tableau* for this dual basic form is

y_1	1	1	1	-1	2
y_2	-1	-3	-1	2	-1
-1	1	0	0	-1	0
	$= y_3$	$= y_4$	$= y_5$	$= y_6$	$= g$

Here, the *columns* are read as equations. For example, column 2 is read

$$(1)(y_1) + (-3)(y_2) + (0)(-1) = y_4,$$

which is the second constraint equation in the dual basic form. A pivot on the first entry in the first row produces Dual Basic Form (3.14), whose dual tableau is

y_3	1	1	1	-1	2
y_2	1	-2	0	1	1
-1	-1	-1	-1	0	-2
	$= y_1$	$= y_4$	$= y_5$	$= y_6$	$= g$

A quick check shows that this second dual tableau could have been obtained by applying the pivoting algorithm (page 80). This is no coincidence. Indeed, *the pivoting algorithm works for dual tableaus exactly as for primal ones.* We omit the proof of this fact; it is similar to the proof of Theorem 3.6. There is also a theorem analogous to Theorem 3.5. It is the following:

THEOREM 3.13. *Consider a dual tableau of size $(m+1) \times (n+1)$. Then we have:*

(1) *The tableau is feasible if and only if each entry in the bottom row (not counting the last one) is nonpositive.*
(2) *Suppose the tableau is feasible. If each entry in the right-hand column (not counting the bottom one) is nonnegative, then the tableau is optimal.*

Thus, the first dual tableau above is not feasible, but the second is feasible and optimal. There is, of course, a dual version of the simplex algorithm:

ALGORITHM 3.4 (THE DUAL SIMPLEX ALGORITHM). *We are given a feasible dual tableau T of size $(m+1) \times (n+1)$.*

(1) *If*
$$t_{i,n+1} \geq 0 \quad for \quad 1 \leq i \leq m,$$
then STOP (the tableau is optimal).

(2) *Choose any k with $1 \leq k \leq m$ such that $t_{k,n+1} < 0$. (The nonbasic variable labeling row k is called the* entering *variable.)*

(3) *If*
$$t_{kj} \geq 0 \quad for \quad 1 \leq j \leq n,$$
STOP (the problem is unbounded.)

(4) *Choose any l with $1 \leq l \leq n$ such that $t_{kl} < 0$ and*
$$\frac{t_{m+1,l}}{t_{kl}} = \min\left\{ \frac{t_{m+1,j}}{t_{kj}} : 1 \leq j \leq n \quad and \quad t_{kj} < 0 \right\}.$$

(The basic variable labeling column l is called the leaving *variable.)*

(5) *Pivot on t_{kl} and replace T by the dual tableau resulting from this pivot. (Entry t_{kl} is called the* pivot entry.)*

(6) *Go to Step (1).*

If this algorithm is modified by Bland's rule, then cycles are impossible, and so the computation always terminates after finitely many iterations. There is, finally, a dual version of the feasibility algorithm which we illustrate by solving the following example:

$$\text{minimize} \quad y_1 + y_2 \qquad\qquad (3.24)$$
$$\text{subject to} \quad y_1 - 2y_2 + y_3 \geq 1$$
$$-y_1 - y_3 \geq -5.$$

A dual basic form is

$$\text{minimize} \quad y_1 + y_2$$
$$\text{subject to} \quad y_4 = y_1 - 2y_2 + y_3 - 1$$
$$y_5 = -y_1 - y_3 + 5.$$

This is infeasible. Construct an auxiliary problem by introducing an auxiliary variable v, an auxiliary objective function $+v$, and auxiliary constraints formed from the original constraints by adding v. The result is

$$\text{minimize} \quad v$$
$$\text{subject to} \quad y_4 = y_1 - 2y_2 + y_3 + v - 1$$
$$y_5 = -y_1 - y_3 + v + 5.$$

The dual tableau for the auxiliary problem is

y_1	1	-1	0
y_2	-2	0	0
y_3	1	-1	0
v	1	1	1
-1	1	-5	0
	$= y_4$	$= y_5$	$= h$

We pivot on the entry in the row labeled by v and in the first column. The reason for this choice is that the entry at the bottom of the first column is positive. If there had been more than one column with positive bottom entry, we would have chosen the largest one. The result of the pivot is

y_1	-1	-2	-1
y_2	2	2	2
y_3	-1	-2	-1
y_4	1	1	1
-1	-1	-6	-1
	$= v$	$= y_5$	$= h$

This is feasible. The dual simplex algorithm allows two choices of pivot. These are the entry in row 1, column 1, and the entry in row 3, column 1. We choose the first of these. The result is

v	-1	2	1
y_2	2	-2	0
y_3	-1	0	0
y_4	1	-1	0
-1	-1	-4	0
	$= y_1$	$= y_5$	$= h$

This is optimal and the auxiliary variable v is nonbasic. To obtain the initial feasible dual tableau for the original problem, we erase the row labeled by v and replace the auxiliary objective function by the original one (written in terms of the nonbasics y_2, y_3, and y_4. The result is

y_2	2	-2	3
y_3	-1	0	-1
y_4	1	-1	1
-1	-1	-4	-1
	$= y_1$	$= y_5$	$= g$

This is feasible but not optimal. The dual simplex algorithm calls for a pivot on the entry in row 2 and column 1. The result is

	2	-2	1
y_2	2	-2	1
y_1	-1	0	1
y_4	1	-1	0
-1	-1	-4	0
	$= y_3$	$= y_5$	$= g$

This is optimal and we read off the solution

$$\min g(\bar{y}) = 0, y_1 = 0, y_2 = 0, y_3 = 1.$$

The reader is asked, in an exercise, to write down a formal statement of the dual feasibility algorithm.

The following theorem summarizes the results about dual problems.

THEOREM 3.14. *A dual linear programming problem which is both feasible and bounded has a solution. Moreover, a solution can be computed by using the dual feasibility algorithm (if necessary), together with the dual simplex algorithm modified by Bland's rule.*

3.5.2. *The Duality Theorem*

Consider a dual/primal pair of problems, and introduce slack and surplus variables so as to obtain a primal and a dual basic form. The tableau and dual tableau corresponding to these basic forms consist of the same $(m + 1) \times (n + 1)$ matrix, and differ only in the labels. This suggests that we combine the two tableaus into one dual/primal tableau. This is done by putting both sets of labels on the matrix. For example, if the primal problem is

$$\text{maximize} \quad -x_1 + 2x_2 - 3x_3 + 4x_4$$

$$\text{subject to} \quad x_3 - x_4 \le 0$$

$$x_1 - 2x_3 \le 1$$

$$2x_2 + x_4 \le 3$$

$$-x_1 + 3x_2 \le 5,$$

then the dual/primal tableau is

	x_1	x_2	x_3	x_4	-1		
y_1	0	0	1	-1	0	=	$-x_5$
y_2	1	0	-2	0	1	=	$-x_6$
y_3	0	2	0	1	3	=	$-x_7$
y_4	-1	3	0	0	5	=	$-x_8$
-1	-1	2	-3	4	0	=	f
	$= y_5$	$= y_6$	$= y_7$	$= y_8$	$= g$		

Given this notation, we can simultaneously carry out pivots on both problems —it is simply a matter of keeping track of both sets of labels. For example, if we pivot on the second entry in the third row of the example, we get

	x_1	x_7	x_3	x_4	-1		
y_1	0	0	1	-1	0	=	$-x_5$
y_2	1	0	-2	0	1	=	$-x_6$
y_6	0	$1/2$	0	$1/2$	$3/2$	=	$-x_2$
y_4	-1	$-3/2$	0	$-3/2$	$1/2$	=	$-x_8$
-1	-1	-1	-3	3	-3	=	f
	$= y_5$	$= y_3$	$= y_7$	$= y_8$	$= g$		

We are now ready for the most important theorem of this chapter.

THEOREM 3.15 (THE DUALITY THEOREM). *If either of a dual/primal pair of problems has a solution, then so does the other; moreover, if they have solutions, the optimal values of the two objective functions are equal to each other.*

PROOF. Suppose first that the primal problem has a solution. Set up the dual/primal tableau and carry out a sequence of pivots which produces an optimal feasible tableau for the primal. This is possible by Theorem 3.12. The final tableau in this sequence has all nonnegative entries in the last column (not counting the bottom row), and all nonpositive entries in the bottom row (not counting the last column). But, then, by Theorem 3.13, the dual tableau is both feasible and optimal. This proves that the dual problem has a solution. Also, the optimal values of both objective functions are equal to the negative of the bottom right-hand entry in the tableau, and thus are equal to each other.

The proof that the primal problem has a solution if the dual does (and that the optimal values are equal) is identical to what we have just done, except for obvious changes.   □

Exercises

(1) Solve

$$\text{minimize} \quad y_1 + y_2 + y_3$$
$$\text{subject to} \quad -y_1 + 2y_2 \geq 1$$
$$-y_2 + 3y_3 \geq 1.$$

(2) Solve the following primal problem and its dual

$$\text{maximize} \quad 5x_1 + x_2 + 5x_3$$
$$\text{subject to} \quad x_1 + 2x_2 + x_3 \leq 4$$
$$x_1 \leq 2$$
$$x_3 \leq 2.$$

(3) Solve the following dual problem and its primal

$$\text{minimize} \quad y_1 - y_2 + y_3$$
$$\text{subject to} \quad -y_1 + y_2 \geq 0$$
$$-y_2 + y_3 \geq -1.$$

(4) Solve the following dual problem and its primal

$$\text{minimize} \quad y_1 - y_2 + y_3$$
$$\text{subject to} \quad 2y_1 - y_2 \geq -1$$
$$-y_1 + y_3 \geq -2.$$

(5) Write down the dual feasibility algorithm.

4
Solving Matrix Games

In this chapter, we apply linear programming to matrix games. The first task is to set up the problem of computing optimal strategies for the row player and column player as a dual/primal pair of linear programming problems. The minimax theorem will then be proved. The rest of the chapter consists of examples.

The connection between game theory and linear programming was first discussed in [Dan51]. Our interest is in solving matrix games by using linear programming methods, that is, in stating the problem of solving a matrix game as a dual/primal pair. It is interesting that, conversely, linear programming problems can be stated as matrix games. See [LR57] for more information about this.

4.1. The Minimax Theorem

Let M be an $m \times n$ matrix game. The *row player's problem* is to compute the row value v_r and an optimal mixed strategy

$$\vec{p} = (p_1, \ldots, p_m).$$

That is, the problem is to find v_r and a vector of probabilities $\vec{p}$ such that v_r is as large as possible and

$$v_r = \min_j \sum_{i=1}^m p_i m_{ij}.$$

We rephrase this problem in an equivalent way (and, also, drop the subscript r from v). The new formulation is: Find v and $\vec{p}$ such that v is as large as possible, subject to the following conditions:

(1) $p_i \geq 0$ for $1 \leq i \leq m$.
(2) $\sum_{i=1}^m p_i = 1$.
(3) $v \leq \sum_{i=1}^m p_i m_{ij}$ for $1 \leq j \leq n$.

To see that this really is equivalent to the first formulation, note that, since v is as large as possible, condition (3) implies that

$$v = \min_j \sum_{i=1}^m p_i m_{ij}.$$

This new formulation is a linear programming problem. It has the disadvantage that the unknown v cannot be assumed to be nonnegative. (In many games, the column player has an advantage; in such games, $v < 0$.) It also is not in the form most convenient for solving. Now, the problem about v not being positive is easily taken care of. According to Exercise (11) on page 51, adding a constant to every entry in M results in adding the same constant to v_r and v_c. If we add a constant so large that the modified matrix has all entries positive, then v_r and v_c will also be positive. Clearly, too, adding a constant to every entry does not change the optimal strategies. Before going further, we assume that the following has been carried out.

PRELIMINARY STEP

Choose a constant c large enough so that $m_{ij} + c > 0$ for all entries m_{ij} in M. Replace M by the result of adding c to every entry.

The plan is to solve the matrix game as modified by this preliminary step. The optimal strategies will be correct for the original matrix; the row value and column values of the original matrix will be those of the modified matrix minus the constant c.

We return to the second formulation of the row player's problem. Let us make a change of variable

$$y_i = p_i/v \text{ for } 1 \leq i \leq m. \tag{4.1}$$

The division by v is legal because of the preliminary step. Now, the condition that the p_i's sum to 1 [condition (2)] implies that

$$\frac{1}{v} = \sum_{i=1}^{m} y_i.$$

Also, condition (3) implies that

$$\sum_{i=1}^{m} m_{ij} y_i \geq 1 \ \text{ for } \ 1 \leq j \leq n.$$

Since maximizing v is equivalent to minimizing $1/v$, we arrive at another (and final) formulation:

$$\text{minimize} \quad y_1 + \cdots + y_m \qquad (4.2)$$

$$\text{subject to} \quad \sum_{i=1}^{m} m_{ij} y_i \geq 1, \ \text{ for } \ 1 \leq j \leq n.$$

Thus, the row player's problem is a dual linear programming problem. It should come as no surprise that the column player's problem is the corresponding primal. The derivation is very similar. First, we can formulate her problem as: Find v and $\vec{q}$ such that v is as small as possible, subject to the following conditions:

(1) $q_j \geq 0$, for $1 \leq j \leq n$.
(2) $\sum_{j=1}^{n} q_j = 1$.
(3) $v \geq \sum_{j=1}^{n} q_j m_{ij}$, for $1 \leq i \leq m$.

The appropriate change of variables is

$$x_j = q_j/v, \ \text{ for } \ 1 \leq j \leq n. \qquad (4.3)$$

The column player's problem becomes

$$\text{maximize} \quad x_1 + \cdots + x_n \qquad (4.4)$$

$$\text{subject to} \quad \sum_{j=1}^{n} m_{ij} x_j \leq 1, \ \text{ for } \ 1 \leq i \leq m.$$

Thus, as predicted, Problems (4.2) and (4.4) form a dual/primal pair. The method of solving a matrix game is now clear. It is:

(1) Carry out the preliminary step (and remember the constant c; it will be needed later).
(2) Set up and solve Problems (4.2) and (4.4).

(3) Compute the column value of the modified (by the preliminary step) game by taking the reciprocal of the maximum value of the primal objective function.

(4) Compute $\vec{p}$ and $\vec{q}$, using the solutions of the linear programming problems and the change of variable equations [(4.1) and (4.3)].

(5) Compute the row value and column value of the original game by subtracting c from the corresponding quantities for the modified game.

It is time to state the following:

THEOREM 4.1 (THE MINIMAX THEOREM). *Let M be an $m \times n$ matrix game. Then, both the row player and column player have optimal mixed strategies. Moreover, the row value and column values are equal.*

PROOF. Problem (4.2) is feasible by Exercise (6). Its objective function is bounded below by zero. By Theorem 3.14, it has a solution. By Theorem 3.15 (the duality theorem), Problem (4.4) also has a solution. Therefore, both players have optimal mixed strategies. Finally, the optimal values of the two objective functions are equal; it follows immediately that the row value and column value are equal. ☐

Let us do an example to illustrate the method. Let

$$M = \begin{pmatrix} -2 & 2 & 1 \\ 0 & -1 & 3 \\ 2 & 1 & -1 \\ -1 & 3 & 0 \end{pmatrix}.$$

This game has no saddle point and there are no dominated rows or columns. We carry out the preliminary step with $c = 3$. The modified matrix is

$$M = \begin{pmatrix} 1 & 5 & 4 \\ 3 & 2 & 6 \\ 5 & 4 & 2 \\ 2 & 6 & 3 \end{pmatrix}.$$

The initial dual/primal tableau for the row player's and column player's problems is

	x_1	x_2	x_3	-1		
y_1	1	5	4	1	=	$-x_4$
y_2	3	2	6	1	=	$-x_5$
y_3	5	4	2	1	=	$-x_6$
y_4	2	6	3	1	=	$-x_7$
-1	1	1	1	0	=	f
	$= y_5$	$= y_6$	$= y_7$	$= g$		

The primal tableau is already feasible, so we carry out the (primal) simplex algorithm. There are three pivots allowed by the algorithm. We choose the entry at row 3, column 1. The result of the pivot is

	x_6	x_2	x_3	-1		
y_1	$-1/5$	$21/5$	$18/5$	$4/5$	$=$	$-x_4$
y_2	$-3/5$	$-2/5$	$24/5$	$2/5$	$=$	$-x_5$
y_5	$1/5$	$4/5$	$2/5$	$1/5$	$=$	$-x_1$
y_4	$-2/5$	$22/5$	$11/5$	$3/5$	$=$	$-x_7$
-1	$-1/5$	$1/5$	$3/5$	$-1/5$	$=$	f
	$= y_3$	$= y_6$	$= y_7$	$= g$		

There are now two possible pivots; the one chosen is the entry at row 4, column 2. After this second pivot, one more pivot gives an optimal tableau

	x_6	x_7	x_5	-1		
y_1	$18/55$	$-54/55$	$-3/10$	$1/11$	$=$	$-x_4$
y_7	$-7/55$	$1/55$	$1/5$	$1/11$	$=$	$-x_3$
y_5	$3/11$	$-2/11$	0	$1/11$	$=$	$-x_1$
y_6	$-3/110$	$12/55$	$-1/10$	$1/11$	$=$	$-x_2$
-1	$-13/110$	$-3/55$	$-1/10$	$-3/11$	$= f$	
	$= y_3$	$= y_4$	$= y_2$	$= g$		

From this, we read off the solutions to the linear programming problems:

$$\max f = \min g = 3/11,$$

$$x_1 = x_2 = x_3 = 1/11,$$

$$y_1 = 0, \quad y_2 = 1/10, \quad y_3 = 13/110, \quad y_4 = 3/55.$$

From the change of variable equations ((4.1) and (4.3)), we see that

$$p_j = (11/3)y_j \quad \text{and} \quad q_i = (11/3)x_i.$$

Thus,

$$\vec{q} = (1/3, 1/3, 1/3) \quad \text{and} \quad \vec{p} = (0, 11/30, 13/30, 1/5).$$

Finally, the value of the original game is the value of the modified game (which is 11/3) minus the constant c from the preliminary step,

$$v(M) = 11/3 - 3 = 2/3.$$

Exercises

(1) Solve the game
$$\begin{pmatrix} -3 & 2 & 0 \\ 1 & -2 & -1 \\ -1 & 0 & 2 \\ 1 & 1 & -3 \end{pmatrix}.$$

(2) Solve the game
$$\begin{pmatrix} 1 & -1 & 1 \\ -1 & 1 & 1 \\ 1 & 1 & -1 \end{pmatrix}.$$

(3) Solve the game
$$\begin{pmatrix} -1 & 2 & -1 & 1 \\ 1 & 0 & 2 & -1 \\ -1 & 1 & -2 & 2 \end{pmatrix}.$$

(4) Solve the game
$$\begin{pmatrix} -2 & 3 & 0 & 1 \\ 1 & -3 & 2 & 0 \\ 0 & -2 & 1 & 3 \\ -1 & 1 & -1 & 2 \end{pmatrix}.$$

(5) Solve the game
$$\begin{pmatrix} 1 & 0 & 0 & 0 \\ 0 & 1 & 0 & 0 \\ 0 & 0 & 1 & 0 \\ 0 & 0 & 0 & 1 \end{pmatrix}.$$

(6) Prove that Problem (4.2) is feasible.

4.2. Some Examples

In this section, we use linear programming to solve some interesting games.

4.2.1. *Scissors-Paper-Stone*

This game was discussed in Chapter 2 (Example 2.1) and the reader was asked to solve it in Exercise (3) on page 63. The three pure strategies play indistinguishable roles in the game and so it is not surprising that the optimal strategy for each player is to play each of them with equal probability: $\vec{p} = (1/3, 1/3, 1/3)$.

We now introduce a more complicated game based on scissors-paper-stone. This game, which we will call *scissors-paper-stone-glass-water*, or

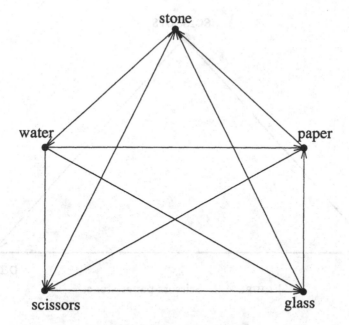

FIGURE 4.1. Scissors-paper-stone-glass-water.

SPSGW, was first discussed in [Wil66]. In SPSGW, the two players simultaneously choose one of the five objects in the name of the game. The winner is determined by looking at Figure 4.1.

In this directed graph, there is an edge between every pair of vertices. If the edge goes from vertex u to vertex v, and if one player chooses u while the other chooses v, then the winner is the one who chooses u. Thus, glass beats stone; stone beats water; scissors beats glass etc. For comparison, the directed graph for ordinary scissors-paper-stone is given in Figure 4.2.

The matrix for SPSGW is easily computed:

$$\begin{pmatrix} 0 & -1 & 1 & 1 & -1 \\ 1 & 0 & 1 & -1 & -1 \\ -1 & -1 & 0 & -1 & 1 \\ -1 & 1 & 1 & 0 & -1 \\ 1 & 1 & -1 & 1 & 0 \end{pmatrix}.$$

Since this is a symmetric game, it could be solved using the method of Chapter 2. The linear programming method was used instead. The value of the game is zero, and the optimal strategy for each player is

$$\vec{p} = (1/9, 1/9, 1/3, 1/9, 1/3).$$

This solution is surprising in at least one way. The third pure strategy seems to be the least favorable for each player. Three of the five outcomes

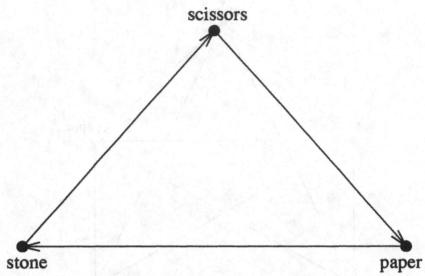

FIGURE 4.2. Scissors-paper-stone.

of playing it are losses, and one of the others is a draw. Nevertheless, the optimal strategy calls for playing it one-third of the time. This illustrates the fact that, even for fairly simple games, the optimal solution may be difficult to arrive at intuitively.

4.2.2. *Three-Finger Morra*

The rules are the same as for two-finger morra, except that each player can hold up one, two, or three fingers (and, at the same time, guess 1, 2, or 3). Thus, each player has $3 \times 3 = 9$ pure strategies. Each of these pure strategies can be designated by an ordered pair (f, p), where f is the number of fingers held up, and p is the prediction. In the matrix for this game, we have labeled the rows and columns by the appropriate designations:

	(1,1)	(1,2)	(1,3)	(2,1)	(2,2)	(2,3)	(3,1)	(3,2)	(3,3)
(1,1)	0	2	2	-3	0	0	-4	0	0
(1,2)	-2	0	0	0	3	3	-4	0	0
(1,3)	-2	0	0	-3	0	0	0	4	4
(2,1)	3	0	3	0	-4	0	0	-5	0
(2,2)	0	-3	0	4	0	4	0	-5	0
(2,3)	0	-3	0	0	-4	0	5	0	5
(3,1)	4	4	0	0	0	-5	0	0	-6
(3,2)	0	0	-4	5	5	0	0	0	-6
(3,3)	0	0	-4	0	0	-5	6	6	0

This game is certainly symmetric. The matrix has no saddle points, and

TABLE 4.1. Solutions for three-finger morra.

Solution	(1,3)	(2,2)	(3,1)
1	5/12	1/3	1/4
2	16/37	12/37	9/37
3	20/47	15/47	12/47
4	25/61	20/61	16/61

there are no dominated rows or columns. It is an example of a symmetric game for which the method of Chapter 2 fails. In fact, none of the nine possible systems of linear equations has any solutions at all. Solving the dual/primal pair of linear programming problems by hand is tedious, but possible. However, they were solved by computer, and four optimal basic solutions were found; these were converted to solutions of the game. It is interesting that they all involve only the three pure strategies $(1,3)$, $(2,2)$, and $(3,1)$. The probabilities of playing these three pure strategies in each of the four solutions are given in Table 4.1.

It is also interesting to note how close together these four solutions are.

4.2.3. *Colonel Blotto's Game*

The version of this game which we will discuss is found in [Dre81]. It is a military game and goes as follows. Colonel Blotto leads an infantry force consisting of four regiments. The enemy force, commanded by General Attila, consists of three regiments. There are two positions which both armies would like to capture (called San Juan Hill and Lookout Mountain). The problem for both commanding officers is to decide how many regiments to send to each position. We assume that a battle between two forces will result in a victory for the one with more regiments, and a draw if they are equal. The payoffs are computed as follows: If a force of r regiments defeats a force of s regiments, then the winner gains $s+1$ (the $+1$ is there because the position is captured by the winner and is regarded as worth 1). Now, the five pure strategies for Colonel Blotto are: $(4,0)$, $(0,4)$, $(3,1)$, $(1,3)$, and $(2,2)$. Here, for example, $(3,1)$ means that three regiments are sent to San Juan Hill and one to Lookout Mountain. Similarly, General Attila has four pure strategies: $(3,0)$, $(0,3)$, $(2,1)$, and $(1,2)$. The payoff matrix is

	(3,0)	(0,3)	(2,1)	(1,2)
(4,0)	4	0	2	1
(0,4)	0	4	1	2
(3,1)	1	−1	3	0
(1,3)	−1	1	0	3
(2,2)	−2	−2	2	2

These entries are easy to compute. For example, if Blotto plays $(3,1)$ and Attila plays $(2,1)$, Blotto wins the battle of San Juan Hill and gains $2+1=3$. The battle of Lookout Mountain is a draw. If Blotto plays $(4,0)$ and Attila plays $(2,1)$, the battle of San Juan Hill is a victory for Blotto, who gains $2+1=3$. The battle of Lookout Mountain is won by Attila, who gains $0+1=1$. Thus, Blotto gains a net payoff of $3-1=2$.

By the linear programming method, we get the following solution:

$$\text{value} = 14/9,$$

$$\vec{p} = (4/9, 4/9, 0, 0, 1/9),$$

$$\vec{q} = (1/30, 7/90, 8/15, 16/45).$$

This says that Blotto should almost always concentrate all his regiments at one of the two positions, but that, one time in nine, he should divide them equally. As for Attila, he should divide his forces most of the time (with probability $8/15 + 16/45 = 8/9$, to be precise), but should occasionally concentrate them all at one position. Thus, the stronger army and the weaker one use dramatically different strategies. There is another curious thing about this solution. Although the two positions are indistinguishable, Attila's pure strategies $(3,0)$ and $(0,3)$ are played with unequal probabilities. The same is true of the pure strategies $(2,1)$ and $(1,2)$. This apparent paradox is removed by the observation that the probability vector obtained from $\vec{q}$ by interchanging q_1 and q_2, and interchanging q_3 and q_4 is also an optimal strategy for General Attila. If we average these two optimal solutions, we get a symmetric optimal solution for Attila: $\vec{q} = (1/18, 1/18, 4/9, 4/9)$. The fact that the value of the game is positive is to be expected since Blotto has one more regiment than Attila; thus, he should have an advantage.

There is a reasonable philosophical objection to the solution above. It is certainly valid in case the "game" is played repeatedly, but this particular military campaign will only occur once. The consequences of an unfortunate combination of pure strategies might be devastating for one of the armies. For example, if Attila plays $(2,1)$ while Blotto plays $(4,0)$, two of Attila's regiments might be rendered unfit for further action for a long time. In the context of the entire war, this might be unacceptable. On the other hand, it is hard to see what a "safe" alternative is for Attila (except to retreat without fighting at all).

4.2.4. *Simple Poker*

Real poker is usually played by more than two people and is difficult to analyze, partly because the variety of possible hands of cards is very great. For this reason, various forms of simplified poker have been studied in order to gain insight into the real game. Our version is played as follows. First, there are two players: San Antonio Rose and Sioux City Sue. Each player *antes* \$1. This means that that amount of money is placed in the middle of the table. The resulting pile of money (or chips representing money) is called the *pot*.

Each player is then dealt a hand consisting of two cards, face down. Each player may look at her own cards, but cannot see the other player's. The deck of cards from which these hands are dealt is highly simplified. It consists of only two kinds of cards: High (abbreviated H), and Low (abbreviated L). Moreover, the deck is very large, so that the probability of dealing an H is always the same as dealing an L. Thus,

$$\Pr(H) = \Pr(L) = 1/2.$$

There are three possible hands: HH,LL,HL. The probabilities of these are easily seen to be

$$\Pr(HH) = 1/4, \quad \Pr(LL) = 1/4, \quad \Pr(HL) = 1/2.$$

In accordance with the traditions of poker, the ranking of the hands is: $HH > LL > HL$. Thus, either pair beats a hand without a pair.

Rose plays first. She has two choices: She can either *bet* by adding \$2 to the pot, or *check* (that is, bet zero). It is now Sue's turn. Her options depend on what Rose did. If Rose bet, Sue can either *see* or *fold* . To see the bet means to match Rose's bet by putting her own \$2 in the pot. To fold means to give up (without having to put any money in), and, in this case, Rose wins the pot. On the other hand, if Rose checks, Sue can either bet \$2 or check. Finally, if Rose checks and Sue bets, then Rose has the option of either seeing the bet or folding (in which case, Sue wins the pot). If neither folds, the two hands are turned over and compared. If they are the same, the pot is divided equally. Otherwise, the player with the higher hand takes the whole pot.

The extensive form of our version of simple poker is large but not unmanageable. To see what Rose's pure strategies are, two things should be kept in mind. First, Rose knows her own hand but not Sue's. Second, if a given pure strategy calls for her to check, then that strategy must also tell her which action to take in case Sue's response is to bet. The first of these considerations tells us that the action to be taken according to a particular strategy can depend on her hand. Since there are three possible hands, it is natural to represent a pure strategy as a 3-tuple, where each entry represents an action for one of the possible hands. To designate these actions, we let "b," "s," "f" stand, respectively, for "bet," "check and see if Sue bets," and "check and fold if Sue bets." There are thus $3^3 = 27$ pure strategies. A typical strategy is "bsf," which means: If Rose's hand is HH, she bets; if it is LL, she checks but will see if Sue bets; and, it is HL, she will check and fold if Sue bets.

As for Sue's pure strategies, note that she knows not only her own hand but also the action taken by Rose. Therefore, there are six situations in which she could find herself (any of three hands times either of two actions taken by Rose). It is therefore natural to represent Sue's pure strategies as 6-tuples. The structure of these 6-tuples is indicated in the following table:

| HH, bet | HH, check | LL, bet | LL, check | HL, bet | HL, check |

We use uppercase letters in designating Sue's possible choices: "B" means "bet," "S" means "see," "C" means "check," and "F" means "fold." A typical pure strategy for Sue is: SBFBSC. The meaning of this is that if Sue is dealt HH and Rose bets, then she sees the bet; if Rose checks, she bets. If Sue is dealt LL and Rose bets, she folds; if Rose checks, she bets. Finally, if Sue is dealt HL and Rose bets, she sees; if Rose checks, she checks.

Since each entry in the 6-tuple can be either of two possibilities (S/F or B/C, depending on what action Rose took), the total number of pure strategies for Sue is $2^6 = 64$. Thus, it appears that the game matrix for simple poker will be 27×64 (taking Rose to be the row player). Common sense (or, rather, poker sense) allows us to eliminate some of the pure strategies for each player. For Rose, we can safely eliminate all strategies which call for her to fold HH. This hand cannot lose. This leaves 18 pure strategies for her. In Sue's case, any strategy which calls for *her* to fold HH can also be eliminated. With a little additional thought, we see that it is safe to eliminate those which call for her to *check* with HH. This reduces the number of pure strategies for Sue to 16. All of them begin with SB. The other way of looking at this is that, if we wrote out the full 27×64 matrix, the rows and columns corresponding to pure strategies we have eliminated would be dominated.

We now must explain how the matrix entries are computed. This is complicated because each entry is an average over all the 9 possible combinations of hands for the two players. To take an example, suppose Rose plays sbs, while Sue plays SBFBFB. We list the nine possible combinations of hands and analyze each:

(1) Each is dealt HH [probability $= (1/4) \times (1/4) = 1/16$]; Rose checks, Sue bets, Rose sees; payoff is 0.

(2) Rose is dealt HH, Sue gets LL [probability $= (1/4) \times (1/4) = 1/16$]; Rose checks, Sue bets, Rose sees; Rose wins $3 (the half of the pot contributed by Sue; the other half was already hers).

(3) Rose is dealt HH, Sue gets HL [probability $= (1/4) \times (1/2) = 1/8$]; Rose checks, Sue bets, Rose sees; Rose wins $3.

(4) Rose is dealt LL, Sue gets HH [probability $= (1/4) \times (1/4) = 1/16$]; Rose bets, Sue sees; Sue wins and so payoff $= -\$3$.

(5) Each is dealt LL [probability $= (1/4) \times (1/4) = 1/16$]; Rose bets, Sue folds; Rose wins $1.

(6) Rose is dealt LL, Sue gets HL [probability $= (1/4) \times (1/2) = 1/8$]; Rose bets, Sue folds; Rose wins $1.

(7) Rose is dealt HL, Sue gets HH [probability $= (1/2) \times (1/4) = 1/8$]; Rose checks, Sue bets, Rose sees; the payoff is $-\$3$.

(8) Rose is dealt HL, Sue gets LL [probability $= (1/2) \times (1/4) = 1/8$]; Rose checks, Sue bets, Rose sees; the payoff is $-\$3$.

(9) Each is dealt HL [probability $= (1/2) \times (1/2) = 1/4$]; Rose checks, Sue bets, Rose sees; the payoff is 0.

The matrix entry in the row labeled sbs, and in the column labeled SBFBFB is the sum of the nine payoffs, each multiplied by the probability of that particular combination occurring:

$$(1/16)(0) + (1/16)(3) + (1/8)(3) + (1/16)(-3) + (1/16)(1) +$$
$$(1/8)(1) + (1/8)(-3) + (1/8)(-3) + (1/4)(0) = -3/16.$$

Computing all 288 entries is a very tedious job, if done by hand. In fact, however, it was done on a computer. Because of space considerations, the matrix will not be shown here. It turns out not to have a saddle point, but there are some dominated rows and columns. After these are eliminated, we are left with a 12×12 matrix. Solving by the linear programming method gives us:

$$\text{value of simple poker is } -0.048.$$

Thus, the game is slightly (about a nickel per hand) favorable to Sue. The dual/primal problem has many solutions. One of these gives the following optimal mixed strategies.

Rose should play

- bsb with probability 0.231
- bsf with probability 0.615
- bff with probability 0.077
- sff with probability 0.077.

Sue should play

- SBSBFB with probability 0.154
- SBSCFC with probability 0.462
- SBFCFB with probability 0.077
- SBFCFC with probability 0.308.

These mixed strategies can be restated in a way which is easier to grasp. Suppose, for example, that Rose is dealt HH. How should she play? Looking at her solution, we see that bsb, bsf, and bff call for her to bet on this hand.

The fourth pure strategy, sff, calls for her to check (and see). The sum of the probabilities of those pure strategies calling for a bet is

$$0.231 + 0.615 + 0.077 = 0.923.$$

Our recommendation to her is, therefore, that she bet with probability 0.923, when dealt HH. Similar calculations are easily made in case she is dealt LL or HL. In summary, our recommendations to Rose are as follows:

- On HH, bet with probability 0.923, check and see with probability 0.077.
- On LL, check and see with probability 0.846, check and fold with probability 0.154.
- On HL, bet with probability 0.231, check and fold with probability 0.769.

Our recommendations to Sue are computed in a similar way. They are as follows:

- On HH and Rose betting, always see.
- On HH and Rose checking, always bet.
- On LL and Rose betting, see with probability 0.616, fold with probability 0.384.
- On LL and Rose checking, bet with probability 0.154, check with probability 0.846.
- On HL and Rose betting, always fold.
- On HL and Rose checking, bet with probability 0.231, check with probability 0.769.

Some of these recommendations are intuitive but some are certainly not. It is to be expected that Rose should occasionally *bluff* with an HL, that is, bet it as if it were a stronger hand. She is told to do this almost one time in four. On the other hand, it would have been difficult to guess that Rose should check HH about one time in 13.

We have mentioned that the dual/primal linear programming problem for this game has many solutions. It is a fascinating fact that if we convert any of the known solutions into recommendations as was done above, we get exactly the same probabilities!

Exercises

(1) Verify that the solution given for SPSGW is correct.

(2) A surprising feature of the solution to SPSGW was discussed earlier in the text, namely, that the third pure strategy is played relatively frequently. Suppose that the row player has no faith in game theory and decides to vary the optimal strategy by playing row 3 with probability less than 1/3, and row 5 more than

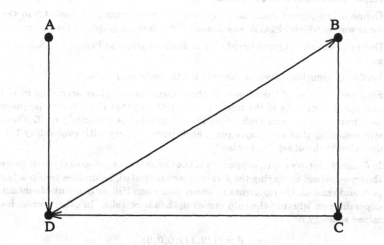

FIGURE 4.3. Directed graph for Exercise (3).

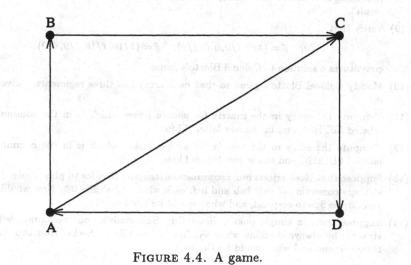

FIGURE 4.4. A game.

1/3 (the other probabilities being unchanged). How would the column player respond and what would the result be?

(3) Define a two-person game based on the directed graph in Figure 4.3 in the same way in which SPSGW was based on its directed graph. Solve the game.

(4) Define a two-person game based on the directed graph in Figure 4.4. Solve the game.

(5) Verify the solutions given in the text for three-finger morra.

(6) Suppose that one of the players in three-finger morra plays according to the first optimal strategy in the table on page 107, but that the other player plays each pure strategy with probability 1/9. Compute the expected payoff. Then, still assuming that one player plays each pure strategy with probability 1/9, how should the other player play?

(7) In Colonel Blotto's game, suppose that the Colonel does not entirely trust game theory—instead of playing the strategy we computed, he decides to flip a fair coin and send all his regiments to either San Juan Hill or Lookout Mountain, depending on whether the coin comes up heads or tails. In other words, his mixed strategy is

$$\vec{p} = (1/2, 1/2, 0, 0, 0).$$

Assuming that Attila knows this, how would you advise him to respond, and what would be the result? Does Attila do better than the value of the game predicts?

(8) Suppose that General Attila ignores our advice and plays the mixed strategy

$$(0, 0, 1/2, 1/2).$$

Assuming that Blotto knows this, how should he play, and what would be the result?

(9) Verify that

$$v = 14/9, \quad \vec{p} = (4/9, 4/9, 0, 0, 1/9), \quad \vec{q} = (1/18, 1/18, 4/9, 4/9)$$

constitutes a solution to Colonel Blotto's game.

(10) Modify Colonel Blotto's game so that each army has three regiments. Solve it.

(11) Compute the entry in the matrix for simple poker which is in the column labeled SBFBFB and in the row labeled bfb.

(12) Compute the entry in the matrix for simple poker which is in the column labeled SBFBSC and in the row labeled bsb.

(13) Suppose that Rose rejects our recommendations and decides to play a mixed strategy consisting of only bsb and bsf, each with probability 0.5. How would you advise Sue to respond, and what would be the result?

(14) Suppose that, in simple poker, Sioux City Sue modifies our recommended strategy by always checking when she has HL and Rose checks. How should Rose respond and what would be the result?

5
Non-Zero-Sum Games

In studying games which are not zero-sum, the distinction between *cooperative* and *noncooperative* games is crucial. There are two types of cooperation among players. The first is the making of binding agreements before play starts about how they will play (that is, coordination of strategies); the second is the making of agreements about sharing payoffs (or about "side payments" from one player to another). The aim of the first type is to increase the payoffs to the cooperating players; the aim of the second is for one player (or group of players) to induce another player to coordinate strategies.

It is important to understand that the agreements made concerning coordination of strategies must be binding. If "double crosses" are possible, then there might as well be no cooperation at all. An interesting example to keep in mind when thinking about these ideas is Prisoner's Dilemma (page 25). In this game, each of the two players (named Bonnie and Clyde) has a choice of two strategies: C (for "cooperate") or D (for "defect"). If they both play D, they both go to prison for five years. If they both play C, they each receive a two-year sentence. If one defects and the other cooperates, the defector gets one year, while the cooperator gets ten. It is clear that if they could play cooperatively (and make a binding agreement), they would both play C. On the other hand, if the game is noncooperative, the best way for each to play is D. We will look at this game in another way in Section 5.2.4. Another interesting point about it is that the payoffs are not transferable. That is, there is no possibility of pooling the payoffs and dividing them in accordance with a previous agreement.

Zero-sum two-person games are noncooperative in nature because cooperation is never to either player's advantage. The total payoff is always zero and cannot be increased by any combination of strategies.

In this chapter, we are interested in games with two players. However, a few theorems and definitions will be stated for an arbitrary number, N, of players. We will study noncooperative games first and then go on to cooperative ones.

5.1. Noncooperative Games

The normal form of a two-person game can be represented as a pair of $m \times n$ matrices (see page 30). A more succinct way of expressing the same thing is as a *bi-matrix*, that is, a matrix of pairs. For example, the bi-matrix for Prisoner's Dilemma is

$$\begin{pmatrix} (-2,-2) & (-10,-1) \\ (-1,-10) & (-5,-5) \end{pmatrix}.$$

In general, if C is an $m \times n$ bi-matrix, then each entry c_{ij} is an ordered pair of numbers. The members of this ordered pair are the payoffs to the two players P_1, P_2 (called, in analogy with the zero-sum case, the row player and column player, respectively), given that the row player plays row i, and the column player plays column j. We will use both the bi-matrix notation (especially for displaying games), and the two-matrix notation (especially for computations).

Other examples follow.

EXAMPLE 5.1 (BATTLE OF THE BUDDIES). Two friends, named Norm and Cliff, have different tastes in evening entertainment. Norm prefers professional wrestling matches, but Cliff likes roller derby. Neither likes to go to his choice of entertainment alone; in fact, each would rather go with the other to the other's choice than go alone to his own. On a numerical scale, each grades going alone to either event as a 0, going to his friend's choice with him as 1, and going with his friend to his own favorite event as 5. We think of these numbers as "happiness ratings."

Regarding this situation as a game, we see that each player has two pure strategies: W (for "wrestling") and R (for "roller derby"). Let us suppose that they independently and simultaneously announce their decisions each evening. This makes the game noncooperative; we will discuss the cooperative variant later. The 2×2 bi-matrix is thus

$$\begin{pmatrix} (5,1) & (0,0) \\ (0,0) & (1,5) \end{pmatrix}$$

We have designated Norm as the row player and Cliff as the column player.

For this game, the advantages of cooperation are not so striking as in the Prisoner's Dilemma game. Of course, both Norm and Cliff are better off if they both choose the same activity—the problem is that they cannot agree on which one to choose. Rational people would resolve this problem by flipping a coin or alternating between the activities, but that would make the game cooperative.

EXAMPLE 5.2 (CHICKEN). Two teenage males with cars meet at a lonely stretch of straight road. They position the cars a mile apart, facing each other, and drive toward each other at a high rate of speed. The cars straddle the center line of the road. If one of the drivers swerves before the other, then the swerver is called "chicken" and loses the respect of his peers. The nonswerver, on the other hand, gains prestige. If both swerve, neither is considered very brave but neither really loses face. If neither swerves, they both die. We assign numerical values, somewhat arbitrarily, to the various outcomes. Death is valued at -10, being chicken is 0, not swerving when the other driver does is worth 5, swerving at the same time as the other is valued at 3.

Our version of this game is taken from [Pou92]. For other versions, see the movies *Rebel Without a Cause* and *American Graffiti*. The bi-matrix for Chicken is

$$\begin{pmatrix} (3,3) & (0,5) \\ (5,0) & (-10,-10) \end{pmatrix}.$$

Cooperating players would surely agree that both should swerve. This conclusion is, however, highly dependent on our choice of payoff numbers. For example, if $(3,3)$ in the bi-matrix is replaced by $(2,2)$, then cooperating players would do better flipping a coin to decide who swerves and who does not.

5.1.1. *Mixed Strategies*

The concept of an equilibrium pair of strategies was introduced in Chapter 1 (see page 24). The strategies to which this definition refers are pure strategies (since mixed strategies had not yet been defined). There is no reason, however, why players in a non-zero-sum game cannot play mixed strategies, and no reason why we cannot extend the definition of an equilibrium pair to include them.

DEFINITION 5.1. Let $\vec{\pi}$ be an N-player game in normal form with strategy sets $X_1, \ldots, X_N$. A *mixed strategy* for player P_i is a probability vector $\vec{p}_i = (p_i(x))_{x \in X_i}$. The entry $p_i(x)$ is interpreted as the probability that P_i plays strategy $x \in X_i$.

For a two-player game the notation is simpler. Let the $m \times n$ matrices A and B be the payoff matrices for the row player and column player, respectively. Then a mixed strategy for the row player is an m-tuple $\vec{p}$ of probabilities; a mixed strategy for the column player is an n-tuple $\vec{q}$ of probabilities.

We will use the symbol M_i to denote the set of all mixed strategies for player P_i.

The *expected payoff* to player P_i due to the mixed strategies $\vec{p}_1, \ldots, \vec{p}_N$, played by players $P_1, \ldots, P_N$, respectively, is

$$\pi_i(\vec{p}_1, \ldots, \vec{p}_N) = \sum (p_1(x_1) \times \cdots \times p_N(x_N))\pi_i(x_1, \ldots, x_N),$$

where the sum is taken over all choices of $x_i \in X_i$, $1 \le i \le N$.

Thus, the expected payoff is the sum of all payoffs due to N-tuples of pure strategies, each weighted according to the probability of that N-tuple being played.

If $N = 2$ and A and B are the $m \times n$ payoff matrices for the row player and column player, respectively, then the expected payoff to the row player due to the mixed strategies $\vec{p}$ and $\vec{q}$ is

$$\pi_1(\vec{p}, \vec{q}) = \sum_{i=1}^{m} \sum_{j=1}^{n} p_i q_j a_{ij}.$$

The expected payoff to the column player is

$$\pi_2(\vec{p}, \vec{q}) = \sum_{i=1}^{m} \sum_{j=1}^{n} p_i q_j b_{ij}.$$

In the game of the Battle of the Buddies, a mixed strategy for Norm is an assignment of probabilities to the two pure strategies W and R. Suppose he decides to flip a fair coin each evening and choose W if it comes up heads, and R if tails. If, meanwhile, Cliff always chooses R, then the expected happiness of Norm is

$$(1/2)(0) + (1/2)(1) = 1/2,$$

while the expected happiness of Cliff is

$$(1/2)(0) + (1/2)(5) = 5/2.$$

On the other hand, if Cliff chooses R with probability 2/3, and W with probability 1/3, then Norm's expected happiness is 7/6, while Cliff's is 11/6.

5.1.2. Maximin Values

Before formally defining equilibrium N-tuples, we discuss a quantity called the "maximin value," which a player can rather easily compute and which gives him or her a pessimistic estimate of how much payoff can be expected. To make the discussion definite, assume that we are dealing with a two-person game with players P_1 and P_2. The maximin value, v_1, for P_1 is computed by making the assumption that the other player will act so as to minimize P_1's payoff. Thus,

$$v_1 = \max_{\vec{p}} \min_{\vec{q}} \pi_1(\vec{p}, \vec{q}),$$

where $\vec{p}$ and $\vec{q}$ range over all mixed strategies for P_1 and P_2, respectively.

Now, unless the game is zero-sum, the assumption that P_2 will play that way is probably false. In fact, P_2 will act so as to maximize her payoff, and this is often not the same as minimizing his. Nevertheless, v_1 gives P_1 a lower bound on his payoff (since he can play so as to guarantee at least that much), and it is easy to compute: We simply regard P_1's payoff matrix as a zero-sum game; its value is then v_1. This is true because, if P_2 plays to minimize P_1's payoff, she is playing to maximize the negatives of those payoffs. This is another way of saying that she is acting as if she were the column player in the matrix game. Consider, for example, the following bi-matrix

$$\begin{pmatrix} (1,1) & (0,1) & (2,0) \\ (1,2) & (-1,-1) & (1,2) \\ (2,-1) & (1,0) & (-1,-1) \end{pmatrix}. \tag{5.1}$$

The payoff matrix for P_1, the row player, is

$$\begin{pmatrix} 1 & 0 & 2 \\ 1 & -1 & 1 \\ 2 & 1 & -1 \end{pmatrix}.$$

As a matrix game, this is easily solved. After removing a dominated row and a dominated column, the remaining 2×2 matrix is solved to give

$$v_1 = 1/2.$$

The optimal strategy (for the matrix game) is to play rows 1 and 3 both with probability 1/2.

The column player's maximin value is then computed by solving the matrix game obtained by transposing her payoff matrix. This transposition is necessary because her payoffs are the actual matrix entries and not their negatives. The value of this transposed game will then be v_2. Here is the matrix we need for the example:

$$\begin{pmatrix} 1 & 2 & -1 \\ 1 & -1 & 0 \\ 0 & 2 & -1 \end{pmatrix}.$$

This is also easily solved to give

$$v_2 = -1/4.$$

The mixed strategy for P_2 which achieves this value is to play column 1 of the bi-matrix with probability 1/4, and column 2 with probability 3/4.

The pessimistic nature of these maximin values is illustrated by the fact that if both players play according to the mixed strategies just computed (which are called the maximin solutions), then they both receive expected payoffs greater than their maximin values. In fact, P_1 actually has an expected payoff of 3/4, while P_2 has an expected payoff of 3/8.

In the special case where the game is zero-sum, the minimax theorem implies that the two maximin values are negatives of each other.

5.1.3. Equilibrium N-tuples of Mixed Strategies

The definition of an equilibrium N-tuple of mixed strategies is

DEFINITION 5.2. Let $\vec{\pi}$ be an N-person game in normal form. An N-tuple of mixed strategies $\vec{q}_1, \ldots, \vec{q}_N$ is an *equilibrium N-tuple* if

$$\pi_i(\vec{q}_1, \ldots, \vec{p}_i, \ldots, \vec{q}_N) \leq \pi_i(\vec{q}_1, \ldots, \vec{q}_N),$$

for all i and for all mixed strategies $\vec{p}_i$ for player P_i.

Thus, if all players but one use the mixed strategies in the N-tuple, but the other departs from it, the one departing suffers (or, at least, does no better).

The following theorem is due to Nash (see [Nas51]; proofs are also to be found in [Owe82] and [Vor77]). We will omit the proof since it uses the Brouwer fixed-point theorem from topology.

THEOREM 5.1. *Let $\vec{\pi}$ be any N-player game. Then there exists at least one equilibrium N-tuple of mixed strategies.*

This result is of fundamental importance. Note that it would be false if we replaced "mixed" by "pure."

The reservations expressed in Chapter 1 about the usefulness of equilibrium N-tuples in solving games are still valid for equilibrium N-tuples of mixed strategies. Nevertheless, they are still useful in some cases. In general, computing equilibrium N-tuples is difficult. We now explore a simple but interesting case where it can be done.

5.1.4. A Graphical Method for Computing Equilibrium Pairs

We illustrate the method by working with the Battle of the Buddies. Since each player has only two pure strategies, the mixed strategies for Norm can all be written in the form $(x, 1-x)$, where $0 \le x \le 1$. Similarly, the mixed strategies for Cliff all have the form $(y, 1-y)$, where $0 \le y \le 1$. The expected payoffs, as functions of x and y, are easy to compute. We have

$$\pi_1(x,y) = 5xy + (1-x)(1-y) = 6xy - x - y + 1,$$

where we have abbreviated $\pi_1((x, 1-x), (y, 1-y))$ by $\pi_1(x,y)$. Similarly,

$$\pi_2(x,y) = xy + 5(1-x)(1-y) = 6xy - 5x - 5y + 5.$$

In order to find an equilibrium pair, we need to find a pair (x^*, y^*) so that $\pi_1(x^*, y^*)$ is a maximum over $\pi_1(x, y^*)$, and $\pi_2(x^*, y^*)$ is a maximum over $\pi_2(x^*, y)$. The idea is to graph together the following two sets:

$$A = \{(x,y) : \pi_1(x,y) \text{ is a maximum over } x, \text{ with } y \text{ fixed}\},$$

and

$$B = \{(x,y) : \pi_2(x,y) \text{ is a maximum over } y, \text{ with } x \text{ fixed}\}.$$

The points of intersection of these sets will be precisely the equilibrium pairs.

First, we rewrite the payoff functions as

$$\pi_1(x,y) = (6y - 1)x - y + 1,$$

and

$$\pi_2(x,y) = (6x - 5)y - 5x + 5.$$

From the formula for $\pi_1(x,y)$, we see that if $y < 1/6$, then $x = 0$ at the maximum (since the coefficient of x is negative); if $y > 1/6$, $x = 1$ at the maximum; and, if $y = 1/6$, any x gives the maximum. For $\pi_2(x,y)$, we see that if $x < 5/6$, $y = 0$ at the maximum; if $x > 5/6$, $y = 1$ at the maximum; and, if $x = 5/6$, any y gives the maximum. The sets A and B are graphed in Figure 5.1. The set A is indicated with a solid line, the set B by a dashed line, and the three equilibrium pairs are circled. They are $((0,1), (0,1))$, $((5/6, 1/6), (1/6, 5/6))$, and $((1,0), (1,0))$. Norm naturally prefers the third of these (since his payoff is 5), and Cliff prefers the first. The second gives each of them 5/6, which is less than each receives at the other two. In fact, 5/6 is the maximin value for both players [by Exercise (3)].

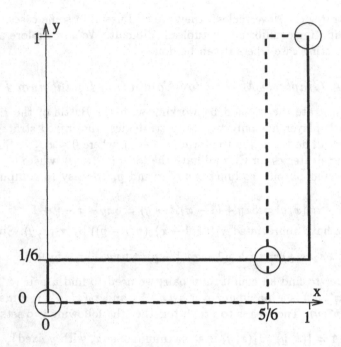

FIGURE 5.1. Equilibrium pairs for Norm and Cliff.

Another example is

$$\begin{pmatrix} (4,-4) & (-1,-1) \\ (0,1) & (1,0) \end{pmatrix}. \tag{5.2}$$

For this game,

$$\pi_1(x,y) = 4xy - x(1-y) + (1-x)(1-y) = (6y-2)x - y + 1,$$

while

$$\pi_2(x,y) = -4xy - x(1-y) + (1-x)y = (-4x+1)y - x.$$

The two sets are shown in Figure 5.2. We see that the only equilibrium pair is $((1/4, 3/4), (1/3, 2/3))$. Playing in accordance with these mixed strategies, the row player has an expected payoff of $2/3$, and the column player has an expected payoff of $-1/4$.

Exercises

(1) In Battle of the Buddies, suppose Cliff plays the mixed strategy $(1/4, 3/4)$. What is the best way for Norm to play in response?

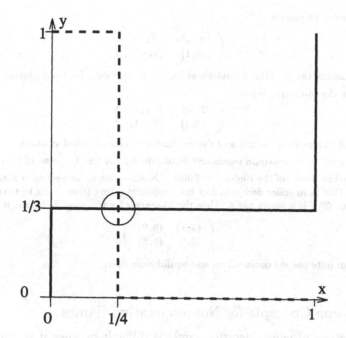

FIGURE 5.2. Equilibrium pair for Bi-matrix (5.2).

(2) For the game of Prisoner's Dilemma, verify that the maximin value for each player is −5, and that $((0,1),(0,1))$ is the only equilibrium pair of mixed strategies.

(3) For the game of Battle of the Buddies, verify that 5/6 is the maximin value for each player.

(4) For the game given by Bi-matrix (5.2), verify that the maximin value for the row player is 2/3, and for the column player is −1.

(5) Given the bi-matrix

$$\begin{pmatrix} (-1,3) & (1,0) \\ (2,-1) & (0,1) \\ (1,1) & (-2,1) \end{pmatrix},$$

compute the maximin values for both players.

(6) Consider the bi-matrix

$$\begin{pmatrix} (-2,3) & (-1,1) & (1,-2) \\ (0,1) & (-1,-2) & (1,1) \\ (2,2) & (2,-1) & (0,0) \end{pmatrix}.$$

Compute the maximin values for both players.

(7) For the bi-matrix

$$\begin{pmatrix} (2,-3) & (-1,3) \\ (0,1) & (1,-2) \end{pmatrix},$$

compute the equilibrium pairs and the maximin values for both players.

(8) For the bi-matrix

$$\begin{pmatrix} (2,-1) & (-1,1) \\ (0,2) & (1,-1) \end{pmatrix},$$

find the maximin values and the equilibrium pairs of mixed strategies.

(9) Compute the maximin values and equilibrium pairs for the game of Chicken.

(10) Modify Battle of the Buddies as follows: Norm is not as interested in wrestling as Cliff is in roller derby; in fact his happiness rating from going to wrestling with Cliff is 2 instead of 5. Thus the bi-matrix for the modified game is

$$\begin{pmatrix} (2,1) & (0,0) \\ (0,0) & (1,5) \end{pmatrix}.$$

Compute the maximin values and equilibrium pairs.

5.2. Solution Concepts for Noncooperative Games

The analysis of noncooperative games is difficult because it is so often dependent on nonmathematical considerations about the players. As a result, most conclusions are subject to honest disagreement. In this respect, these games may be closer to real life than are other parts of game theory! We illustrate this by discussing several examples. Before doing so, we will enlarge our stock of analytic tools by making some definitions. The first of these is the following:

DEFINITION 5.3. Let $\vec{\pi}$ be a two-person game. The two-dimensional set

$$\Pi = \{(\pi_1(\vec{p}, \vec{q}), \pi_2(\vec{p}, \vec{q})) : \vec{p} \in M_1, \vec{q} \in M_2\}$$

is called the *noncooperative payoff region*. The points in Π are called *payoff pairs*.

The payoff region can be plotted in a Cartesian coordinate system, where the horizontal axis represents π_1 and the vertical axis represents π_2. The idea is to consider all possible pairs of mixed strategies and plot the corresponding pair of payoffs for each. One thing that can be read from a drawing of a payoff region is the relation of *dominance* between payoff pairs. This is defined by the following:

DEFINITION 5.4. Let (u, v) and (u', v') be two payoff pairs. Then (u, v) *dominates* (u', v') if

$$u \geq u' \text{ and } v \geq v'.$$

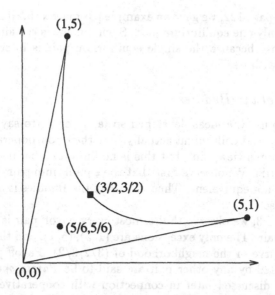

FIGURE 5.3. Payoff region for Battle of the Buddies.

In the payoff region, the dominating payoff pair is above and to the right ("northeast") of the dominated one.

For Battle of the Buddies, the payoff region is shown in Figure 5.3. It is bounded by two line segments and the curve from $(1, 5)$ to $(5, 1)$. (This curve is a piece of a parabola.) The three (distinct) entries in the bimatrix are the coordinates of points in the region resulting from playing pure strategies. The three equilibrium pairs of mixed strategies result in the points marked with small circles.

Our second definition is the following:

DEFINITION 5.5. Let both $(\vec{p}, \vec{q})$ and $(\vec{r}, \vec{s})$ be equilibrium pairs of mixed strategies for a two-person game. Then

(1) they are said to be *interchangeable* if $(\vec{p}, \vec{s})$ and $(\vec{r}, \vec{q})$ are also equilibrium pairs;

(2) they are said to be *equivalent* if

$$\pi_i(\vec{p}, \vec{q}) = \pi_i(\vec{r}, \vec{s}),$$

for $i = 1, 2$.

Theorem 2.6 shows that in a zero-sum game, all equilibrium pairs of mixed strategies are interchangeable and equivalent. This is not always true of non-zero-sum games. It is sometimes said that a game in which any two equilibrium pairs are interchangeable and equivalent is *solvable in the*

Nash sense. On page 122, we gave an example [Bi-matrix (5.2)] of a game in which there is only one equilibrium pair. Such a game is certainly solvable in the Nash sense because the single equilibrium pair is interchangeable and equivalent with itself.

5.2.1. *Battle of the Buddies*

Let us try to use the ideas developed so far in order to say something about how Norm and Cliff might actually play their noncooperative game. We will take Norm's viewpoint, but this is no limitation because the game is really symmetric. We observe first that the equilibrium pairs are neither interchangeable nor equivalent. Thus, Battle of the Buddies is not solvable in the Nash sense.

From Figure 5.3, we observe that almost every payoff pair is dominated by some other pair. The only exceptions are $(1,5)$, $(5,1)$, and the points on the bounding curve in the neighborhood of $(3/2, 3/2)$. Payoff pairs which are not dominated by any other pair are said to be *Pareto optimal*. This concept will be discussed later in connection with cooperative games. It is clear that rational cooperating players would never play so that their payoff pair is not Pareto optimal.

We list some possible mixed strategies and comment on them. The first three listed are Norm's strategies in the equilibrium pairs of mixed strategies. These are always of great interest because it is natural to think that only an equilibrium pair has the stability to persist indefinitely.

(1) Norm might play the pure strategy W if he thinks that Cliff will eventually decide that he will never play anything else; based on this decision, Cliff's only rational move is to play W himself. The success of this scenario requires a difference in degree of stubbornness between Norm and Cliff. Otherwise, by symmetry, Cliff might persist in playing R. This combination of W by Norm and R by Cliff results in the lowest possible payoff for both.

(2) Norm should probably not play $(5/6, 1/6)$ (his strategy in the second equilibrium pair). Cliff could gain $5/6$ playing either W or R; if he happens to choose R, Norm's payoff would drop to $1/6$.

(3) The strategy R is an interesting choice for Norm. If Cliff believed this pattern of play would continue, he could respond with R. The result would be excellent for Cliff, and not bad for Norm. Symmetry, however, comes into this. If R is a good idea for Norm, is not W a good idea for Cliff? That would be a rather comic turn of events, but the possibility of it does cast doubt on the idea.

(4) The maximin value for each player is $5/6$ [from Exercise (3) of the previous section]. Norm can assure himself of at least that by playing $(1/6, 5/6)$. In fact, if he plays that way, his payoff is the

same no matter how Cliff plays. Perhaps this is the best he can do.

The conclusion that we just came to is not very satisfying; it is impossible to think of it as definitive. Suppose, for example, that Norm takes our advice and plays $(1/6, 5/6)$. If Cliff sees what his friend is doing, he can, by playing R, gain an expected payoff of $25/6$, far greater than Norm's payoff.

5.2.2. *Prisoner's Dilemma*

In Exercise (2) of the previous section, the reader was asked to verify that the maximin value for each player in Prisoner's Dilemma is -5, and that there is only one equilibrium pair of mixed strategies: $((0, 1), (0, 1))$. The payoff to both players is then -5 for this equilibrium pair. As observed before, this game, like any game with a single equilibrium pair, is solvable in the Nash sense. Now if the row player plays D (the second row), the column player has no reasonable choice; he must play D also. It is therefore highly likely that the payoff pair (D, D) would indeed be played. This is true even though it is not Pareto optimal [since (D, D) is dominated by (C, C)]. Note also that the maximin solutions are the same as the equilibrium pair.

Our conclusion about Prisoner's Dilemma is hard to accept because we want the two prisoners to act more like decent and intelligent human beings. If they do so, the payoff pair will surely be $(-2, -2)$. The trouble is that the game is presented to us as noncooperative. The cooperative variant of it is different and has a different solution. There is a third way to look at this which we discuss shortly.

5.2.3. *Another Game*

The game given in Bi-matrix (5.2) was shown to have only one equilibrium pair of mixed strategies: $((1/4, 3/4), (1/3, 2/3))$. It is thus solvable in the Nash sense. In Exercise (4) of the previous section, the reader was asked to verify that the maximin value for the row player is $2/3$, and for the column player, -1. Now, is it reasonable for the row player to play $(1/4, 3/4)$ (his mixed strategy in the equilibrium pair)? If he does so, the column player gains $-1/4$ playing either column. However, the row player's expected payoff is 1 if the column player plays column 1, and only $1/2$ if she plays column 2. This latter payoff is less than the row player can guarantee himself by playing according to the maximin solution. It is easy to verify that if both play according to the maximin solutions, the column player's actual payoff is $-1/6$, which is greater than the payoff of $-1/4$ from the equilibrium pair. It is therefore likely that if the row player does play $(1/4, 3/4)$, the column player would indeed play the second column in order to force him to change to the maximin solution.

5.2.4. *Supergames*

Let us think about Prisoner's Dilemma in a different way. First of all, there are many conflict situations in which there is a dilemma similar to the one facing the prisoners. These range from ones involving personal relationships between two people to others involving great military questions which may determine the survival of civilization. The features common to all these games are:

- Both players do well if they cooperate with each other.
- If one player plays the cooperative strategy while the other betrays her by playing the defecting strategy, then the defector does very well and the cooperator does badly.
- Neither player really trusts the other.

We discuss two examples. First, there is the H-Bomb Game. In the period immediately following World War II, both the United States and the Soviet Union had the scientific and industrial resources to develop the fusion bomb. It was known that such a weapon would be immensely more destructive than even the atomic bombs which the United States had just dropped on two Japanese cities. For each of the countries, the choice was between building the H-bomb (*defecting* from the wartime alliance between them) or not building it (continuing to *cooperate*). For each country (but more for the Soviet Union), the cost of defecting would be very great. The resources devoted to the project would be unavailable for improving the living standards of the people; also, the societies would continue to be distorted by the dominant role played by military considerations. Furthermore, if both countries defected, there would be a real possibility of a nuclear war which could throw the human race back to the Stone Age. On the other hand, if one country built the bomb and the other did not, the defector would have such an enormous superiority in military force that the very sovereignty of the cooperator would be at risk. Of course, both countries chose defection as the solution to the game.

For a less momentous example, consider two people, Robert and Francesca, who are romantically involved. Each has the choice of either remaining faithful to the other (*cooperating*), or going out with someone else on the sly (*defecting*). If one defects and the other cooperates, the defector is happy (for a while, at least). After all, variety is the spice of life. On the other hand, the cooperator feels betrayed and miserable. If both defect, they each feel fairly good for a while, but the level of tension between them detracts from their joy, and the probability of the relationship breaking up is high. This would make both of them very sad. If this Dating Game is viewed as noncooperative, they would probably both defect. On the other hand, if they trust each other enough, the game becomes cooperative and they will probably both remain faithful. Thus we see that game theory is relevant to romance.

In general, we say that a bi-matrix game is *of Prisoner's Dilemma type* if it has the form

$$\left(\begin{matrix} (a,a) & (b,c) \\ (c,b) & (d,d) \end{matrix} \right), \tag{5.3}$$

where

$$c > a > d > b \text{ and } a > (b+c)/2.$$

These inequalities say that mutual cooperation (row 1, column 1) is better than mutual defection (row 2, column 2) and that, if you fear your opponent is going to defect, it is best to defect also. The last inequality says that, if the players can trust each other, the best way for both to play is to cooperate. Any bi-matrix game which satisfies these conditions will be referred to as Prisoner's Dilemma. Another example is

$$\left(\begin{matrix} (3,3) & (0,5) \\ (5,0) & (1,1) \end{matrix} \right).$$

Some of the conflict situations modeled by Prisoner's Dilemma are such that they can only happen once; others can occur repeatedly.

Our discussion of Prisoner's Dilemma and other noncooperative games is correct if we assume that the game is played only once, or that, if played repeatedly, both players decide on a mixed strategy to which they stick. Suppose, however, that we think about a *supergame* consisting of a certain large number of repetitions of a noncooperative game. Each of the separate plays of the game is called an *iteration* and a player may use different strategies in different iterations. In particular, a player's strategy may depend on previous moves. One way in which this idea could be used is for one player to estimate her opponent's mixed strategy by keeping track of the frequency with which he chooses his various pure strategies. She could then play so as to do the best possible against this (estimated) mixed strategy. For example, if the opponent in the first 100 iterations has played his first pure strategy 27 times, his second pure strategy 39 times, his third never, and his fourth 34 times, then she could estimate his mixed strategy as $(.27, .39, 0, .34)$.

The supergame is just as noncooperative as the original one and so cooperation in the true sense is still not allowed, but a form of communication is now possible. To make this clearer, imagine 500 iterations of Prisoner's Dilemma. The extensive form of this supergame is enormous, but really rather simple in structure. Each nonterminal vertex has exactly two children. (A tree with this property is called *binary*.) We can picture it as follows. Start with the tree for Prisoner's Dilemma. Use each of its terminal vertices as the root of another copy of the tree for Prisoner's Dilemma. Repeat this process as many times as there are iterations in the supergame. There are some information sets in the supergame tree because a player does not know his opponent's move in the same iteration. However, he does know the moves made in previous iterations. Such information could be used in many different ways. For example, a player could use a strategy in which he "signals" his willingness to cooperate by playing C several times in a row. If the other player catches on, they would both play C and do well. If the other player refuses to begin cooperating, then the first player can begin playing D (perhaps sending the cooperation signal again later). Of course, if the opponent never plays anything but D, our player does poorly in those iterations in which he plays C. The point is, however, that there are relatively few of these in comparison to the total number of iterations.

In general, suppose we consider a supergame consisting of M iterations of a certain game (called the *base game*). There are two kinds of supergame strategies. There are those which take account of moves in previous iterations, and those which do not. We call the first kind *adaptive* and the second kind *forgetful*. For example, if the base game is Prisoner's Dilemma, the following are forgetful strategies:

- Always cooperate.
- Always defect.
- Cooperate with probability .5, defect with probability .5.

Here is an adaptive strategy for the same game: Cooperate on the first iteration; after j iterations have been played, compute p_j to be the proportion of the times in these first j iterations in which the opponent has cooperated; in iteration $j+1$, randomly choose to cooperate with probability p_j, and to defect with probability $1-p_j$. In this strategy, the opponent is rewarded for frequently cooperating and punished for frequently defecting.

Another way of looking at supergame strategies is as follows. If $1 \leq j \leq M$, the strategy to be used in the jth iteration can, in general, be a function of the moves made in the first $j - 1$ iterations. A supergame strategy is forgetful if the strategy used at the jth iteration is actually independent of the previous moves.

Some fanciful names have been given to supergame strategies ([Sag93]). The Golden Rule is the strategy of always cooperating. The Iron Rule is the strategy of never cooperating. Tit-for-Tat is the strategy of doing whatever your opponent did on the previous iteration (and cooperating on the first one). In recent years, tournaments have been held in which various supergame strategies have competed against one another. These strategies were submitted by game theorists and the tournaments were organized in round-robin fashion so that each strategy was pitted against every other strategy in a supergame with Prisoner's Dilemma as the base game. The total payoff was added up for each strategy. The consistent winner was Tit-for-Tat. This is interesting partly because of the simplicity of the strategy, and also because it is one which people seem to use instinctively in many conflict situations. Even though Tit-for-Tat won the tournaments by accumulating the highest total payoff against the other strategies, it can never gain more payoff in any single supergame! [See Exercise (8).] The book [Axe84] has much discussion on this subject.

There has been a great deal of interest in these supergames on the part of people interested in a variety of fields, including sociology and biology. Explanations have been sought for the evolution of cooperative behavior in situations where it is difficult to explain in any other way. Some intriguing examples are discussed in [Axe84]. These include both biological evolution and evolution of behaviors in human societies. In particular, the problem is discussed of how a population of individuals acting completely selfishly can evolve into a state in which cooperation is the rule.

Exercises

(1) Analyze the game given in Exercise (8) of the previous section.

(2) Sketch the payoff region for Prisoner's Dilemma.

(3) For the game given in Bi-matrix (5.2), prove that $(1,0)$ is a Pareto optimal payoff pair.

(4) Analyze the modified Battle of the Buddies game described in Exercise (10) of the previous section.

(5) Suppose that you have played 137 iterations of three-finger morra against the same opponent. You notice that he has played $(1,1)$ 41 times, $(2,2)$ 65 times, and $(3,3)$ the rest of the times. Use this data to estimate his mixed strategy. How would you play in response?

(6) Sketch the tree for the supergame consisting of two iterations of Prisoner's Dilemma. Be sure to include information sets and payoffs.

(7) Consider a supergame consisting of M iterations of Prisoner's Dilemma. Prove that the only equilibrium pair for this supergame is the Iron Rule vs. the Iron Rule.

(8) Suppose that a supergame with base game Prisoner's Dilemma (Bi-matrix (5.3)) is being played. One of the players is using Tit-for-Tat as her strategy. Prove that the total payoff accumulated by that player is less than or equal to the total accumulated by her opponent. Prove also that the difference in these total payoffs is at most $c - b$.

5.3. Cooperative Games

In the theory discussed in this section, the players are allowed to make binding agreements about which strategies to play. They do not share payoffs, and there are no side payments from one to another. In many cases, the restriction against sharing payoffs is not due to the rules of the game, but arises from the nature of the payoffs. In Battle of the Buddies, it does not make sense for one of the players to give the other some of his "happiness ratings." In Prisoner's Dilemma, the law requires that each prisoner serve his or her own time. The other prisoner cannot take on the burden. In the next chapter, we will study games in which payoffs can be shared.

Let us consider Battle of the Buddies as a cooperative game. There is an obvious way in which any two cooperating players would surely decide to play the game. Each evening, they would flip a fair coin and both go to the wrestling match if it comes up heads, and both go to the roller derby if tails. In this way, the expected payoff of each would be 3, far higher than they could expect in the noncooperative variant. In Prisoner's Dilemma, the obvious way to play cooperatively is for the players to make a binding agreement with each other to cooperate. In Chicken, the players could agree that they both swerve, or could agree to flip a coin to decide who swerves and who does not. In each of these examples a *joint strategy* is used. The definition is as follows:

DEFINITION 5.6. Let $\bar{\pi}$ be a two-person game with $m \times n$ payoff matrices A and B. A *joint strategy* is an $m \times n$ probability matrix $P = (p_{ij})$. Thus,

$$p_{ij} \geq 0, \quad \text{for } 1 \leq i \leq m, 1 \leq j \leq n,$$

and

$$\sum_{i=1}^{m} \sum_{j=1}^{n} p_{ij} = 1.$$

Thus, a joint strategy assigns a probability to each pair of pure strategies. The expected payoff to the row player due to the joint strategy P is

$$\pi_1(P) = \sum_{i=1}^{m} \sum_{j=1}^{n} p_{ij} a_{ij}.$$

The expected payoff $\pi_2(P)$ to the column player is the same except that a_{ij} is replaced by b_{ij}.

For example, consider the bi-matrix

$$\begin{pmatrix} (2,0) & (-1,1) & (0,3) \\ (-2,-1) & (3,-1) & (0,2) \end{pmatrix}. \tag{5.4}$$

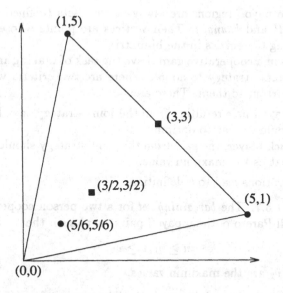

FIGURE 5.4. Cooperative payoff region for Battle of the Buddies.

There are six pairs of pure strategies. Suppose that the players agree to play according to the joint strategy described in the matrix

$$\begin{pmatrix} 1/8 & 0 & 1/3 \\ 1/4 & 5/24 & 1/12 \end{pmatrix}. \tag{5.5}$$

Thus, they will play the pair of pure strategies (row 2, column 1) with probability 1/4, and the pair of pure strategies (row 1, column 3) with probability 1/3. Under this joint strategy, the expected payoff for the row player is

$$(1/8)(2) + (0)(-1) + (1/3)(0) + (1/4)(-2) + (5/24)(3) + (1/12)(0) = 3/8,$$

and for the column player it is 17/24.

The *cooperative* payoff region is the set

$$\{(\pi_1(P), \pi_2(P)) : P \text{ is a joint strategy}\}.$$

It is a larger set than its noncooperative counterpart [according to Exercise (10)]. Thus, there are more payoff pairs obtainable by the players if they cooperate. Figure 5.4 shows the cooperative payoff region for Battle of the Buddies. It should be compared with Figure 5.3.

Cooperative payoff regions are always *convex* sets (defined soon); they are also *closed*[1] and *bounded*. Their vertices are points whose coordinate pairs are among the entries in the bi-matrix.

The players in a cooperative game have the task of making an agreement about which joint strategy to adopt. There are two criteria which would surely be important to them. These are:

- The payoff pair resulting from the joint strategy they have agreed on should be Pareto optimal.
- For each player, the gain from the joint strategy should be at least as great as the maximin value.

These considerations lead to a definition.

DEFINITION 5.7. The *bargaining set* for a two-person cooperative game is the set of all Pareto optimal payoff pairs (u, v) such that

$$u \geq v_1, v \geq v_2,$$

where v_1 and v_2 are the maximin values.

In Figure 5.4, the bargaining set is the line segment from $(1,5)$ to $(5,1)$. The problem for the players is then to agree on a payoff pair in the bargaining set. In Battle of the Buddies, the symmetry of the game strongly suggests the payoff pair $(3,3)$, the midpoint of the line segment. Figure 5.5 shows the cooperative payoff region for Bi-matrix (5.4). The maximin values are $v_1 = 0, v_2 = 2$. Thus the bargaining set is the line segment from $(0,3)$ to $(3/4, 2)$. The row player prefers a payoff pair as far to the right as possible, while the column player prefers one as far up as possible. They negotiate in order to agree on one between the extremes.

5.3.1. *Nash Bargaining Axioms*

The theory which we now present was developed by Nash (see [Nas50]). It is an attempt to establish a fair method of deciding which payoff pair in the bargaining set should be agreed on. The idea is to prove the existence of an *arbitration procedure* Ψ which, when presented with a payoff region P and a *status quo* point $(u_0, v_0) \in P$, will produce a payoff pair (the *arbitration pair*) which is fair to both players. The status quo point is usually the pair of maximin values. (We will mention other choices later.) It is the pair of payoffs which the players must accept if they reject the payoff pair suggested by the arbitration procedure. The arbitration procedure Ψ may be thought of as taking the place of a neutral human arbiter who is called in to settle a conflict. It is required to satisfy the six *Nash axioms*,

---

[1] Roughly speaking, this means they contain their bounding line segments. See [Apo74] or [PM91] for details.

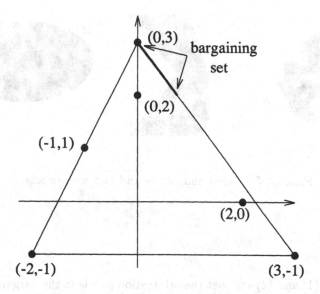

FIGURE 5.5. Payoff region for Bi-matrix (5.4).

which may be thought of as the principles of fairness and consistency which might guide a human arbiter. It will not only be proved that an arbitration procedure exists but that it is unique. This fact lends much credibility to the idea.

We denote the arbitration pair corresponding to payoff region P and status quo point (u_0, v_0) by $\Psi(P, (u_0, v_0)) = (u^*, v^*)$. With this notation, the Nash axioms are:

(1) *[Individual Rationality]* $u^* \geq u_0$ and $v^* \geq v_0$.
(2) *[Pareto Optimality]* (u^*, v^*) is Pareto optimal.
(3) *[Feasibility]* $(u^*, v^*) \in P$.
(4) *[Independence of Irrelevant Alternatives]* If P' is a payoff region contained in P and both (u_0, v_0) and (u^*, v^*) are in P', then

$$\Psi(P', (u_0, v_0)) = (u^*, v^*).$$

(5) *[Invariance Under Linear Transformations]* Suppose P' is obtained from P by the linear transformation

$$u' = au + b, \quad v' = cv + d \text{ where } a, c > 0.$$

Then

$$\Psi(P', (au_0 + b, cv_0 + d)) = (au^* + b, cv^* + d).$$

(6) *[Symmetry]* Suppose that P is symmetric (that is, $(u, v) \in P$ if and only if $(v, u) \in P$), and that $u_0 = v_0$. Then $u^* = v^*$.

FIGURE 5.6. One nonconvex and two convex sets.

Axioms (1) and (2) say that the arbitration pair is in the bargaining set. Axiom (3) simply says that the arbitration pair can be achieved. Axiom (4) says that if a different game has a smaller payoff region and the same status quo point and if this smaller payoff region contains the arbitration pair (u^*, v^*), then the arbitration procedure should suggest the same payoff pair for the other game. Axiom (5) says, in part, that if there is a change in the units in which the payoffs are computed, then there is no essential change in the arbitration pair. Axiom (6) says that if the players have symmetric roles as far as both the payoff region and the status quo point are concerned, then they should gain the same payoff.

5.3.2. Convex Sets

Before stating and proving the theorem that there is a unique arbitration procedure satisfying Nash's axioms, we must discuss a small part of the theory of convexity. The book [Val64] treats this subject in detail and is recommended to the reader who wants to learn more about it.

DEFINITION 5.8. A subset S of $\Re^n$ is said to be *convex* if, for every $\vec{x}$ and $\vec{y}$ in S, and every number t with $0 \le t \le 1$, we have

$$t\vec{x} + (1-t)\vec{y} \in S.$$

In words, this definition says that S is convex if every line segment whose end-points are in S lies entirely in S. In Figure 5.6, two of the sets are convex and the other is not.

In order to study how a convex set is built up from a smaller set, we make the following definitions:

DEFINITION 5.9. Let $F = \{\vec{x}_1, \ldots, \vec{x}_k\}$ be a finite subset of $\Re^n$. Then

$$\sum_{i=1}^{k} t_i \vec{x}_i$$

is a *convex combination* of F whenever $t_1, \ldots, t_k$ are nonnegative numbers whose sum is 1.

We see, by induction, that if S is convex, then any convex combination of points of S is in S.

DEFINITION 5.10. Let A be any subset of $\Re^n$. The *convex hull* of A, denoted co(A), is defined to be the set of all convex combinations of finite subsets of A.

The reader is asked to prove [in Exercise (12)] that co(A) is actually convex. It follows that every convex set which contains A also contains co(A).

For example, a triangle is the convex hull of its vertices; the second convex set in Figure 5.6 is the convex hull of its six vertices; the first convex set in Figure 5.6 is the convex hull of the ellipse which bounds it.

The coefficients t_i in the definition of convex combination have the properties of probabilities. In fact, we have the following:

THEOREM 5.2. *Let $\vec{\pi}$ be a two-person game with $m \times n$ payoff bi-matrix C. Then the cooperative payoff region is the convex hull of the set of points in $\Re^2$ whose coordinates are the entries in the bi-matrix.*

PROOF. If P is a joint strategy, then the corresponding payoff pair is

$$(\pi_1(P), \pi_2(P)) = \sum_{i=1}^{m} \sum_{j=1}^{n} p_{ij} c_{ij},$$

which is in the convex hull of

$$\{c_{ij} : 1 \le i \le m, 1 \le j \le n\}.$$

Conversely, any point of the convex hull is a payoff pair. $\square$

Nash's Axiom (6) referred to symmetric payoff regions. We have

DEFINITION 5.11. A subset S of $\Re^2$ is said to be *symmetric* if (v, u) is in S whenever (u, v) is in S.

Thus, a set is symmetric if it is identical to its reflection in the diagonal $x = y$. Figure 5.7 shows two symmetric sets, one convex and the other not.

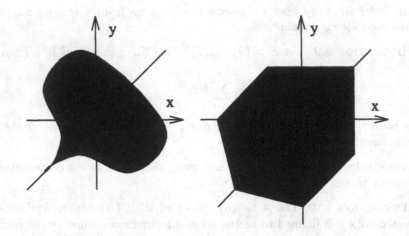

FIGURE 5.7. Two symmetric sets.

DEFINITION 5.12. *Let A be a subset of $\Re^2$. The symmetric convex hull is defined to be the convex hull of the set*

$$A \cup \{(v, u) : (u, v) \in A\}.$$

It is denoted sco(A).

Thus, the symmetric convex hull is formed by reflecting A in the diagonal and then forming the convex hull of the union of A and its reflection. The reader is asked to prove [in Exercise (13)] that the result of this process is actually symmetric. The following is needed later:

LEMMA 5.3. *Let A be a subset of $\Re^2$ and k be a number such that*

$$u + v \leq k,$$

for every point (u, v) in A. Then the same inequality holds for every point in the symmetric convex hull of A.

PROOF. The inequality certainly holds for every point in the reflection of A in the diagonal. Thus, it holds for every point of the union of A and this reflection. The set of *all* points in $\Re^2$ for which the inequality holds is convex. Thus, it contains sco(A). $\square$

5.3.3. *Nash's Theorem*

THEOREM 5.4. *There exists a unique arbitration procedure Ψ satisfying Nash's axioms.*

PROOF. First, we describe how to compute $\Psi(P, (u_0, v_0))$; then we prove that Nash's axioms are satisfied; then, finally, we show that there is only one such arbitration procedure. In computing Ψ, there are two major cases, and three subcases of the second of these.

Case (i). There exists $(u, v) \in P$ such that $u > u_0$ and $v > v_0$. Let K be the set of all (u, v) satisfying these conditions. Then define

$$g(u, v) = (u - u_0)(v - v_0), \text{ for } (u, v) \in K.$$

Let $(u^*, v^*) \in K$ be the point at which $g(u, v)$ attains its maximum value. [The point (u^*, v^*) exists and is unique according to Lemma 5.5 proved later.] Define

$$\Psi(P, (u_0, v_0)) = (u^*, v^*).$$

Case (ii). No $(u, v) \in P$ exists such that $u > u_0$ and $v > v_0$. Consider the following three subcases.

Case (iia). There exists $(u_0, v) \in P$ with $v > v_0$.

Case (iib). There exists $(u, v_0) \in P$ with $u > u_0$.

Case (iic). Neither (iia) nor (iib) is true.

The first thing to notice about these subcases is that (iia) and (iib) cannot both be true. For, suppose that they are and define

$$(u', v') = (1/2)(u_0, v) + (1/2)(u, v_0).$$

Then (u', v') is in P (by convexity) and satisfies the condition of Case (i). Since Case (i) does not hold, this is a contradiction and thus (iia) and (iib) cannot both hold. Now we define $\Psi(P, (u_0, v_0))$ in each of these subcases.

In Case (iia), let v^* be the largest v for which (u_0, v) is in P^2; then define $\Psi(P, (u_0, v_0)) = (u_0, v^*)$.

In Case (iib), let u^* be the largest u for which (u, v_0) is in P; then define $\Psi(P, (u_0, v_0)) = (u^*, v_0)$.

In Case (iic), let $\Psi(P, (u_0, v_0)) = (u_0, v_0)$.

Our arbitration procedure is now defined in all possible cases. We now prove that the Nash axioms hold. Axioms (1) and (3) are obvious in all the cases. Suppose that Axiom (2) is not true. Then there exists $(u, v) \in P$ which dominates (u^*, v^*) and is different from it. Now, in Case (i), we have

$$(u - u_0) \geq (u^* - u_0), (v - v_0) \geq (v^* - v_0),$$

and at least one of these inequalities is strict [since $(u, v) \neq (u^*, v^*)$]. Thus,

$$g(u, v) > g(u^*, v^*).$$

This is a contradiction. In Case (iia), we must have $u^* = u_0 = u$, since Case (iib) does not hold. Thus, $v > v^*$. But this contradicts the definition of v^*. Case (iib) is dealt with in a similar way. In Case (iic), $(u^*, v^*) = (u_0, v_0)$. If

---

$^2 v^*$ exists because P is closed and bounded.

$u > u_0$, then Case (iib) holds; if $v > v_0$, then Case (iia) holds. Since neither of these does hold, we have again reached a contradiction. We conclude that Axiom (2) holds.

We now show that Axiom (4) holds. In Case (i), the maximum value of g over $K \cap P'$ is less than or equal to its maximum over K. But (u^*, v^*) belongs to P' and so the two maxima are equal. Thus

$$\Psi(P', (u_0, v_0)) = \Psi(P, (u_0, v_0)).$$

Cases (iia) and (iib) are similar; Case (iic) is easy.

To verify Axiom (5), suppose first that Case (i) holds. Then Case (i) also holds for payoff region P' with status quo point $(au_0 + b, cv_0 + d)$. Also

$$(u' - (au_0 + b))(v' - (cv_0 + d)) = ac(u - u_0)(v - v_0).$$

Since $a, c > 0$, the maximum of the left-hand side of the equation above is attained at $(au^* + b, cv^* + d)$. Thus the axiom holds in Case (i). The reasoning in the other cases is similar.

Finally, we come to Axiom (6). Suppose that $u^* \neq v^*$. By the symmetry condition, $(v^*, u^*) \in P$. In Case (i), we have

$$g(v^*, u^*) = g(u^*, v^*).$$

By Lemma 5.5, g attains its maximum only at one point, and so this is a contradiction. Cases (iia) and (iib) cannot hold since, if one held, so would the other (by symmetry). As proved above, this is impossible. Case (iic) is obvious.

It remains to prove that Ψ is unique. Suppose that there is another arbitration procedure $\overline{\Psi}$ which satisfies the Nash axioms. Since they are different, there is a payoff region P and a status quo point $(u_0, v_0) \in P$ such that

$$(\overline{u}, \overline{v}) = \overline{\Psi}(P, (u_0, v_0)) \neq \Psi(P, (u_0, v_0)) = (u^*, v^*).$$

Suppose Case (i) holds. Then $u^* > u_0$ and $v^* > v_0$. We define

$$u' = \frac{u - u_0}{u^* - u_0}, v' = \frac{v - v_0}{v^* - v_0}.$$

This linear change of variables takes (u_0, v_0) into $(0, 0)$ and (u^*, v^*) into $(1, 1)$. Thus, by Axiom (5),

$$\Psi(P', (0, 0)) = (1, 1).$$

Also by Axiom (5),

$$\overline{\Psi}(P', (0, 0)) \neq (1, 1).$$

We now verify that if $(u', v') \in P'$, then

$$u' + v' \leq 2.$$

Suppose this is not true. By the convexity of P',

$$t(u', v') + (1 - t)(1, 1) \in P', 0 \le t \le 1.$$

Now define, for $1 \le t \le 1$,

$$h(t) = g(t(u', v') + (1 - t)(1, 1)) = (tu' + (1 - t))(tv' + (1 - t)).$$

Then $h(0) = 1$ and the derivative of $h(t)$ is

$$h'(t) = 2tu'v' + (1 - 2t)(u' + v') - 2(1 - t),$$

so that

$$h'(0) = u' + v' - 2 > 0.$$

Thus, there exists a small positive t such $h(t) > 1$. But this contradicts the definition of (u^*, v^*), since $g(1, 1) = 1$.

Now let $\hat{P}$ be the symmetric convex hull of P'. Then, by Lemma 5.3, $s + t \le 2$ for all $(s, t) \in \hat{P}$, and, therefore, if $(a, a) \in \hat{P}$, $a \le 1$. Since $\hat{P}$ is symmetric, it follows from Axiom (6) that

$$\overline{\Psi}(\hat{P}, (0, 0)) = (1, 1),$$

since, otherwise, there would exist a point (a, a) in $\hat{P}$ with $a > 1$. But, then, by Axiom (4), applied to $\overline{\Psi}$,

$$\overline{\Psi}(P', (0, 0)) = (1, 1).$$

This is a contradiction which proves uniqueness in Case (i).

This leaves Cases (iia), (iib), and (iic). Now, (iic) is easy by Axiom (1). Cases (iia) and (iib) are similar so we only consider the first of these. Since we are not in Case (i), we see, from Axiom (1), that $\overline{u} = u_0 = u^*$. Since both (u^*, v^*) and $(\overline{u}, \overline{v})$ are Pareto optimal, $\overline{v} = v^*$. This contradicts the assumption that $(\overline{u}, \overline{v})$ and (u^*, v^*) are different. $\square$

LEMMA 5.5. *Let P be a payoff region and $(u_0, v_0) \in P$. Suppose that there exists a point $(u, v) \in P$ with*

$$u > u_0, v > v_0,$$

and let K be the set of all $(u, v) \in P$ satisfying these inequalities. Define, on K,

$$g(u, v) = (u - u_0)(v - v_0).$$

Then g attains its maximum on K at one and only one point.

PROOF. The set

$$K' = \{(u,v) \in P : u \geq u_0, v \geq v_0\}$$

is a closed bounded set containing K. By a theorem of mathematical analysis (see [Apo74] or [PM91], for example), the function g, being continuous, attains its maximum on K'. Clearly, the maximum of g over K' is the same as its maximum over K.

It remains to prove that the maximum is attained only once. Suppose it is attained at two different points, (u_1, v_1) and (u_2, v_2). We let

$$M = \max g(u,v) = g(u_1, v_1) = g(u_2, v_2).$$

Now either

$$u_1 > u_2, \qquad v_1 < v_2,$$

or

$$u_1 < u_2, \qquad v_1 > v_2.$$

We carry out the proof assuming the first possibility. The proof with the second possibility assumed is similar. By the convexity of P,

$$(u_3, v_3) = (1/2)(u_1, v_1) + (1/2)(u_2, v_2) \in P.$$

We compute

$$
\begin{aligned}
g(u_3, v_3) \quad &= \quad \left(\frac{u_1 + u_2}{2} - u_0\right)\left(\frac{v_1 + v_2}{2} - v_0\right) \\
&= \quad \left(\frac{u_1 - u_0}{2} + \frac{u_2 - u_0}{2}\right)\left(\frac{v_1 - v_0}{2} + \frac{v_2 - v_0}{2}\right) \\
&= \quad (1/4)[(u_1 - u_0)(v_1 - v_0) + (u_2 - u_0)(v_2 - v_0) \\
&\quad + \quad (u_1 - u_0)(v_2 - v_0) + (u_2 - u_0)(v_1 - v_0)] \\
&= \quad (1/4)[2M + 2M + (u_1 - u_0)(v_2 - v_0) \\
&\quad + \quad (u_2 - u_0)(v_1 - v_0) - (u_1 - u_0)(v_1 - v_0) \\
&\quad - \quad (u_2 - u_0)(v_2 - v_0)] \\
&= \quad M + (1/4)[(v_2 - v_0)(u_1 - u_2) + (v_1 - v_0)(u_2 - u_1)] \\
&= \quad M + (1/4)[(u_1 - u_2)(v_2 - v_1)] \\
&> \quad M.
\end{aligned}
$$

This is a contradiction because M is the maximum. $\square$

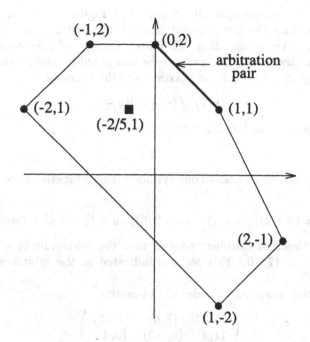

FIGURE 5.8. Cooperative payoff region for Bi-matrix (5.6).

5.3.4. *Computing Arbitration Pairs*

The proof of Theorem 5.4 actually contains an algorithm for computing the arbitration pair. We illustrate the computation with some examples. First, let us look at Battle of the Buddies. The payoff region, which is shown in Figure 5.4, is a symmetric set. The pair of maximin values is $(5/6, 5/6)$. Taking this pair to be (u_0, v_0), we see that Axiom (6) applies. Thus the arbitration pair is of the form (a, a). Since (a, a) must be Pareto optimal [by Axiom (2)], we see that $a = 3$. Thus, we get the point which common sense led us to guess.

For another example, consider the cooperative game given by the bi-matrix

$$\begin{pmatrix} (2, -1) & (-2, 1) & (1, 1) \\ (-1, 2) & (0, 2) & (1, -2) \end{pmatrix}. \tag{5.6}$$

The maximin values are easily computed to be $v_1 = -2/5, v_2 = 1$. The payoff region is shown in Figure 5.8. The pair $(-2/5, 1)$ is indicated with a small square.

Taking $(u_0, v_0) = (-2/5, 1)$, we see that the arbitration pair is to be found among the points in the payoff region which dominate $(-2/5, 1)$, and which are Pareto optimal. A glance at Figure 5.8 shows that these points constitute the line segment from $(1,1)$ to $(0,2)$ (indicated with a heavy line). We see also that, in the terminology of the proof, Case (i) holds. Thus, the arbitration pair can be computed by finding the point on this line segment at which the maximum of the function

$$g(u,v) = (u - u_0)(v - v_0)$$

occurs. Now, on this line segment,

$$v = -u + 2,$$

and so the problem reduces to one of maximizing a function of one variable. We have

$$g(u,v) = (u + 2/5)(v - 1) = (u + 2/5)(-u + 1) = -u^2 + 3u/5 + 2/5.$$

By the methods of calculus, we find that the maximum is attained for $u = 3/10, v = 17/10$. This point is indicated as the arbitration pair in Figure 5.8.

For another example, consider the bi-matrix

$$\begin{pmatrix} (5,1) & (7,4) & (1,10) \\ (1,1) & (9,-2) & (5,1) \end{pmatrix}. \tag{5.7}$$

The payoff region is shown in Figure 5.9.

The maximin values are easily computed to be $v_1 = 3$ and $v_2 = 1$. From the figure, we see that the bargaining set is the union of two line segments. These are the one from $(3,8)$ to $(7,4)$, and the one from $(7,4)$ to $(8,1)$. The bargaining set is indicated with a heavy line. With $(u_0, v_0) = (3,1)$, we are clearly in Case (i). Thus we must maximize

$$g(u,v) = (u - 3)(v - 1)$$

over the bargaining set. At the three points $(3,8)$, $(7,4)$, and $(8,1)$, we have

$$g(3,8) = 0, g(7,4) = 12, g(8,1) = 0.$$

On the upper line segment, $v = -u + 11$ and so

$$g(u,v) = (u - 3)(-u + 11 - 1) = (u - 3)(-u + 10) = -u^2 + 13u - 30.$$

Setting the derivative with respect to u equal to zero gives $u = 13/2$ and $v = 9/2$. This point is in the bargaining set and gives us

$$g(13/2, 9/2) = 49/4.$$

On the lower line segment, $v = -3u + 25$ and so

$$g(u,v) = (u - 3)(-3u + 25 - 1) = (u - 3)(-3u + 24) = -3u^2 + 33u - 72.$$

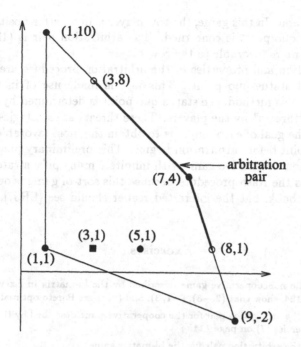

FIGURE 5.9. Payoff region for Bi-matrix (5.7).

Setting the derivative with respect to u equal to zero gives $u = 11/2$ and $v = 17/2$. This point is, however, on the continuation of the line segment beyond $(7, 4)$ and is thus outside the bargaining set. Since $49/4 > 12$, the arbitration pair is $(13/2, 9/2)$.

5.3.5. *Remarks*

The method discussed just now, with the status quo point taken to be the pair of maximin values, is called the *Shapley procedure*.

It is interesting to examine the effect of the status quo point on the arbitration pair. The arbitration pair for Bi-matrix (5.6) is $(3/10, 17/10)$. This seems much more favorable to the column player than to the row player; for both, the maximum possible payoff is 2. The arbitration pair gives the row player only 15 percent of this maximum, but gives the column player 85 percent. This difference is, in part at least, a result of the fact that the status quo point is more favorable to the column player. A verbal description of the situation would be that the column player is in a stronger bargaining position. If the negotiations collapse, she still gets a good payoff, but the row player does not. Bi-matrix (5.7) also illustrates this phenomenon. In this game, the row player is in a better position as far as the status quo point is concerned. The arbitration pair is $(13/2, 9/2)$, which is also more favorable to the row player.

The definition and properties of the arbitration procedure are valid for any choice of status quo point. This fact is made use of in the *Nash procedure*. In this method, the status quo point is determined by means of preliminary "threats" by the players. These threats are strategies (pure or mixed) and the goal of each player is to obtain the most favorable possible status quo point before arbitration begins. This preliminary phase can be viewed as a noncooperative game with infinitely many pure strategies. We do not discuss the Nash procedure because this sort of game is outside the scope of this book, but the interested reader should see [LR57], [Tho84], or [Owe82].

Exercises

(1) For the noncooperative game described by the bi-matrix in Exercise (7) on page 124, show that $(2, -3)$, $(-1, 3)$, and $(0, 1)$ are Pareto optimal.

(2) Find the arbitration pair for the cooperative game described by the bi-matrix in Exercise (7) on page 124.

(3) Find the arbitration pair for the bi-matrix game

$$\begin{pmatrix} (-1, -1) & (4, 0) \\ (0, 4) & (-1, -1) \end{pmatrix}.$$

(4) Find the arbitration pair for the bi-matrix game

$$\begin{pmatrix} (-1, 1) & (0, 0) \\ (1, -1) & (0, 1) \\ (-1, -1) & (1, 1) \end{pmatrix}.$$

(5) Find the arbitration pair for the modified Battle of the Buddies game described in Exercise (10) on page 124.

(6) Find the arbitration pair for Chicken..

(7) Modify Chicken by changing the payoff to the nonswerver from 3 to 2. Compute the arbitration pair for the modified game and compare this answer with the arbitration pair for the original Chicken. How does this innocent-seeming change in the payoff numbers affect the "optimal" way of playing the game?

(8) Find the arbitration pair for the game given in Bi-matrix (5.1).

(9) Find the arbitration pair for the bi-matrix game

$$\begin{pmatrix} (-1/2, 0) & (-1/2, -4) \\ (1, 2) & (-2, 4) \\ (4, -4) & (-1/2, 0) \end{pmatrix}.$$

(10) Prove that the noncooperative payoff region for a two-person game is a subset of the cooperative payoff region.

(11) Let P be the probability matrix (5.5) on page 133. Do there exist mixed strategies $\vec{p}$ and $\vec{q}$ for the two players such that

$$p_i q_j = P_{ij} \quad \text{for all } i, j?$$

(12) Prove that the convex hull of a set is convex.

(13) Prove that the symmetric convex hull of a set is symmetric.

6
N-Person Cooperative Games

In the previous chapter, we studied cooperative games in which the players could coordinate their strategies but could not share payoffs. In the games considered in this chapter, the players are permitted to cooperate fully. Sets of players (called *coalitions*) cannot only make binding agreements about joint strategies, but can also agree to pool their individual payoffs and then redistribute the total in a specified way. In order for this latter kind of cooperation to be possible, we must assume that the payoffs are in some transferable quantity, such as money. The number of players is not explicitly restricted to be greater than two, but might as well be—the case of two players turns out either to be trivial or to have been covered in our earlier work.

The theory developed in this chapter applies both to zero-sum and non-zero-sum games. A large part of it goes all the way back to [vNM44], that is, to the beginning of game theory.

6.1. Coalitions

The idea of a coalition plays an essential role in the theory we are about to develop. As indicated above, a coalition is simply a subset of the set of players which forms in order to coordinate strategies and to agree on how the total payoff is to be divided among its members. As in our earlier discussion of cooperative games, it must be understood that the agreements

TABLE 6.1. A 3-player game.

Strategy triples	Payoff vectors
$(1,1,1)$	$(-2,1,2)$
$(1,1,2)$	$(1,1,-1)$
$(1,2,1)$	$(0,-1,2)$
$(1,2,2)$	$(-1,2,0)$
$(2,1,1)$	$(1,-1,1)$
$(2,1,2)$	$(0,0,1)$
$(2,2,1)$	$(1,0,0)$
$(2,2,2)$	$(1,2,-2)$

which the members of the coalition make among themselves are absolutely binding.

We introduce some notation. The set consisting of all N players is denoted $\mathcal{P}$. A coalition is denoted by an uppercase script letter: $\mathcal{S}, \mathcal{T}, \mathcal{U}$, etc. Given a coalition $\mathcal{S} \subseteq \mathcal{P}$, the *counter-coalition* to $\mathcal{S}$ is

$$\mathcal{S}^c = \mathcal{P} - \mathcal{S} = \{P \in \mathcal{P} : P \notin \mathcal{S}\}.$$

Let us consider a 3-player example. Its normal form is given in Table 6.1. Each player has two strategies, denoted 1 and 2. Thus, there are eight combinations of strategies and eight corresponding 3-tuples of payoffs.

In this game, $\mathcal{P} = \{P_1, P_2, P_3\}$. There are eight coalitions. Three of them contain only one player each. These are

$$\{P_1\}, \{P_2\}, \{P_3\}.$$

Then, there are three of two players each, namely,

$$\{P_1, P_2\}, \{P_1, P_3\}, \{P_2, P_3\}.$$

Finally, $\mathcal{P}$, consisting of all the players, is called the *grand coalition*, and the counter-coalition to $\mathcal{P}$ is $\emptyset$, the *empty coalition*. In general, in a game with N players, there are 2^N coalitions.

6.1.1. *The Characteristic Function*

If a coalition $\mathcal{S}$ has formed, it is natural to think of the game as a contest between two "players," these being the coalition $\mathcal{S}$ and the counter-coalition $\mathcal{S}^c$. This two-"person" game is non-cooperative. Indeed, if the coalition and the counter-coalition cooperated, the grand coalition would form (in place of $\mathcal{S}$).

Suppose, then, that we are dealing with an N-player game for which

$$\mathcal{P} = \{P_1, \ldots, P_N\},$$

and for which the strategy set for player P_i is denoted by X_i. Let $\mathcal{S} \subseteq \mathcal{P}$ be a coalition. We assume, for now, that $\mathcal{S}$ is not empty and does not contain all the players. The pure joint strategies available to this coalition are the members of the Cartesian product of those X_i's for which $P_i \in \mathcal{S}$. Similarly, the pure joint strategies available to the counter-coalition are the members of the Cartesian product of those X_i's for which $P_i \notin \mathcal{S}$. The game between the coalition $\mathcal{S}$ and the counter-coalition $\mathcal{S}^c$ is a bi-matrix game. The rows correspond to the pure joint strategies available to $\mathcal{S}$, and the columns to those available to $\mathcal{S}^c$. An entry in the bi-matrix is a pair whose first member is the sum of the payoffs to the players in the coalition, and whose second member is the sum of the payoffs to the players in the counter-coalition (given that both "players" play according to the pure joint strategies labeling the row and column in which the entry appears).

In the game given by Table 6.1, let us consider the coalition $\mathcal{S} = \{P_1, P_3\}$. Then $\mathcal{S}^c = \{P_2\}$. The coalition has four pure joint strategies available to it. These may be designated: $(1,1)$, $(1,2)$, $(2,1)$, and $(2,2)$, where, for example, $(2,1)$ means that P_1 plays 2, and P_3 plays 1. The counter-coalition has only two pure strategies: 1 and 2. The 4×2 bi-matrix is

	1	2
$(1,1)$	$(0,1)$	$(2,-1)$
$(1,2)$	$(0,1)$	$(-1,2)$
$(2,1)$	$(2,-1)$	$(1,0)$
$(2,2)$	$(1,0)$	$(-1,2)$

$$(6.1)$$

Here, for example, the pair in row 1 and column 2 is $(2,-1)$ because the payoff 3-tuple corresponding to strategy combination $(1,2,1)$ is $(0,-1,2)$. Then the sum of payoffs to the players in the coalition is $0 + 2 = 2$, and the payoff to the counter-coalition is -1.

The maximin value for the coalition (computed from the bi-matrix) is called the *characteristic function* of $\mathcal{S}$ and is denoted $\nu(\mathcal{S})$. As explained in our previous discussion (page 119) of maximin values, the members of $\mathcal{S}$ are guaranteed to be able to choose a joint strategy with which they can gain a total payoff of at least $\nu(\mathcal{S})$. The characteristic function has as its domain the set of all coalitions, and measures the values of these coalitions. Let us do some computing for the example in Table 6.1. We know that $\nu(\{P_1, P_3\})$ is the value of the matrix game whose entries are the first terms in the entries in Bi-matrix (6.1). This matrix is

$$\begin{pmatrix} 0 & 2 \\ 0 & -1 \\ 2 & 1 \\ 1 & -1 \end{pmatrix}.$$

Two of the rows are dominated. The resulting 2×2 matrix game is easily solved to give
$$\nu(\{P_1, P_3\}) = 4/3.$$
The characteristic function of the counter-coalition, $\{P_2\}$, is the value of
$$\begin{pmatrix} 1 & 1 & -1 & 0 \\ -1 & 2 & 0 & 2 \end{pmatrix}.$$
Two columns are dominated. The resulting 2×2 matrix game is solved to give
$$\nu(\{P_2\}) = -1/3.$$
Computing in a similar way, we have
$$\nu(\{P_1, P_2\}) = 1, \quad \nu(\{P_3\}) = 0, \quad \nu(\{P_2, P_3\}) = 3/4, \quad \nu(\{P_1\}) = 1/4.$$
The value of the characteristic function for the grand coalition is simply the largest total payoff which the set of all players can achieve. It is easily seen that
$$\nu(\mathcal{P}) = 1.$$
Finally, by definition, the characteristic function of the empty coalition is
$$\nu(\emptyset) = 0.$$

By examining the values of the characteristic function, we can speculate about which coalitions are likely to form. Since P_1 does better playing on his own than do the other two, it might be that P_2 and P_3 would bid against each other to try to entice P_1 into a coalition. In exchange for his cooperation, P_1 would demand a large share of the total payoff to the coalition he joins. He would certainly ask for more than 1/4, since he can get that much on his own. On the other hand, if he demands too much, P_2 and P_3 might join together, exclude P_1, and gain a total of 3/4.

There is an interesting theorem about the characteristic function. It says that "in union, there is strength."

THEOREM 6.1 (SUPERADDITIVITY). *Let $\mathcal{S}$ and $\mathcal{T}$ be disjoint coalitions. Then*
$$\nu(\mathcal{S} \cup \mathcal{T}) \geq \nu(\mathcal{S}) + \nu(\mathcal{T}).$$

PROOF. By definition of the characteristic function, there is a joint strategy for $\mathcal{S}$ such that the total payoff to the members of $\mathcal{S}$ is at least $\nu(\mathcal{S})$. A similar statement holds for $\mathcal{T}$. Since $\mathcal{S}$ and $\mathcal{T}$ are disjoint, it makes sense for each member of $\mathcal{S}$ and each member of $\mathcal{T}$ to play in accordance with these maximin solutions. The result is a joint strategy for the union of the two coalitions such that the total payoff is guaranteed to be at least
$$\nu(\mathcal{S}) + \nu(\mathcal{T}).$$

The maximin value for the union might be even larger. □

The example we studied earlier shows that the inequality in the theorem may be strict. For instance,

$$\nu(\{P_1, P_3\}) = 4/3 > 1/4 = 1/4 + 0 = \nu(\{P_1\}) + \nu(\{P_3\}).$$

The first of the following two corollaries can be proved using the theorem and mathematical induction. The second corollary is a special case of the first.

COROLLARY 6.2. *If $S_1, \ldots, S_k$ are pairwise disjoint coalitions (that is, any two are disjoint), then*

$$\nu\left(\bigcup_{i=1}^{k} S_i\right) \geq \sum_{i=1}^{k} \nu(S_i).$$

COROLLARY 6.3. *For any N-person game,*

$$\nu(\mathcal{P}) \geq \sum_{i=1}^{N} \nu(\{P_i\}).$$

As far as the formation of coalitions is concerned, a game can be analyzed using only the characteristic function. This suggests the following definition.

DEFINITION 6.1. A game in *characteristic function form* consists of a set

$$\mathcal{P} = \{P_1, \ldots, P_N\}$$

of players, together with a function ν, defined for all subsets of $\mathcal{P}$, such that

$$\nu(\emptyset) = 0,$$

and such that superadditivity holds. That is,

$$\nu(S \cup T) \geq \nu(S) + \nu(T),$$

whenever S and T are disjoint coalitions of the players.

As an abbreviation, we will use the single symbol ν to designate a game in characteristic function form.

6.1.2. *Essential and Inessential Games*

We can single out a class of games which are trivial as far as coalitions are concerned. In fact, they have the property that there is no reason to prefer any one coalition over any other.

DEFINITION 6.2. An N-person game ν in characteristic function form is said to be *inessential* if

$$\nu(\mathcal{P}) = \sum_{i=1}^{N} \nu(\{P_i\}).$$

A game which is not inessential is said to be *essential.*

In other words, a game is inessential if the inequality in Corollary 6.3 is actually an equality. In fact, for such a game, a stronger statement is true.

THEOREM 6.4. *Let S be any coalition of the players in an inessential game. Then*

$$\nu(\mathcal{S}) = \sum_{P \in \mathcal{S}} \nu(\{P\}).$$

PROOF. Suppose not. Then, by Corollary 6.2,

$$\nu(\mathcal{S}) > \sum_{P \in \mathcal{S}} \nu(\{P\}),$$

and so, by superadditivity,

$$\nu(\mathcal{P}) \geq \nu(\mathcal{S}) + \nu(\mathcal{S}^c) > \sum_{i=1}^{N} \nu(\{P_i\}),$$

which contradicts the definition of an inessential game. □

Thus, in an inessential game, there is no reason for a coalition actually to form—cooperation does not result in a greater total payoff.

The fact that a game is inessential does not make it unimportant. It simply means that there is no reason for a coalition to form. The following theorem shows that there are many inessential games which are, nevertheless, important.

THEOREM 6.5. *A two-person game which is zero-sum in its normal form is inessential in its characteristic function form.*

PROOF. The minimax theorem says that $\nu(\{P_1\})$ and $\nu(\{P_2\})$ are negatives of each other, and thus their sum is zero. But, also, $\nu(\mathcal{P}) = 0$, by the zero-sum property.

Thus,

$$\nu(\mathcal{P}) = \nu(\{P_1\}) + \nu(\{P_2\}).$$

$\square$

In the context of the previous chapter, an inessential two-person game is one for which there is no advantage to be gained by cooperating. Zero-sum games with more than two players can be essential. See Exercise (5).

Exercises

(1) Verify the values of the characteristic function for the game shown in Table 6.1.

(2) Compute the characteristic function form of the game shown in Table 1.1, page 26.

(3) Compute the characteristic function (for all coalitions) for the game shown in Table 1.2, page 29.

(4) Show that the following (non-zero-sum) bi-matrix game is inessential:

$$\begin{pmatrix} (0,0) & (1,-2) \\ (-1,1) & (1,-1) \end{pmatrix}.$$

(5) The 3-person game of Couples is played as follows. Each player chooses one of the other two players; these choices are made simultaneously. If a couple forms (for example, if P_2 chooses P_3, and P_3 chooses P_2), then each member of that couple receives a payoff of $1/2$, while the person not in the couple receives -1. If no couple forms (for example, if P_1 chooses P_2, P_2 chooses P_3, and P_3 chooses P_1), then each receives a payoff of zero. Prove that Couples is zero-sum and essential.

(6) Let $\mathcal{P} = \{P_1, P_2, P_3, P_4\}$ be a set of players. Let a, b, c, d, e, f be nonnegative numbers such that

$$\begin{aligned} a + d &\leq 1 \\ b + e &\leq 1 \\ c + f &\leq 1. \end{aligned}$$

Define ν by

$$\nu(\emptyset) = 0, \quad \nu(\mathcal{P}) = 1,$$

$$\nu(\text{any 3-person coalition}) = 1,$$

$$\nu(\text{any 1-person coalition}) = 0,$$

$$\nu(\{P_1, P_2\}) = a, \quad \nu(\{P_1, P_3\}) = b, \quad \nu(\{P_1, P_4\}) = c,$$

$$\nu(\{P_3, P_4\}) = d, \quad \nu(\{P_2, P_4\}) = e, \quad \nu(\{P_2, P_3\}) = f.$$

Prove that ν is a game in characteristic function form.

6.2. Imputations

Suppose that a coalition forms in an N-person game. The problem we now wish to study is that of the final distribution of the payoffs. It is this which, presumably, the players themselves would be most interested in. Indeed, a player considering whether to join a given coalition would want to know how much she gains from doing so. Now, the amounts going to the various players form an N-tuple $\vec{x}$ of numbers. We will argue that there are two conditions which such an N-tuple must satisfy in order that it have any chance of actually occurring in the game. These are *individual rationality* and *collective rationality*. An N-tuple of payments to the players which satisfies both these conditions is called an *imputation*. After the formal definition, we will try to justify the appropriateness of the two conditions.

DEFINITION 6.3. Let ν be an N-person game in characteristic function form with players

$$\mathcal{P} = \{P_1, \ldots, P_N\}.$$

An N-tuple $\vec{x}$ of real numbers is said to be an *imputation* if both the following conditions hold:

- (*Individual Rationality*) For all players P_i,

$$x_i \geq \nu(\{P_i\}).$$

- (*Collective Rationality*) We have

$$\sum_{i=1}^{N} x_i = \nu(\mathcal{P}).$$

The condition of individual rationality is easy to motivate. If

$$x_i < \nu(\{P_i\}),$$

for some i, then no coalition giving P_i only the amount x_i would ever form—P_i would do better going on his own.

As for collective rationality, we first argue that

$$\sum_{i=1}^{N} x_i \geq \nu(\mathcal{P}). \tag{6.2}$$

Suppose that this inequality is false. We would then have

$$\beta = \nu(\mathcal{P}) - \sum_{i=1}^{N} x_i > 0.$$

Thus, the players could form the grand coalition and distribute the total payoff, $\nu(\mathcal{P})$, in accordance with

$$x_i' = x_i + \beta/N,$$

thus giving every player more. Hence, if $\vec{x}$ is to have a chance of actually occurring, (6.2) must be true.

We then argue that

$$\sum_{i=1}^{N} x_i \leq \nu(\mathcal{P}). \tag{6.3}$$

To see this, suppose that $\vec{x}$ actually occurs. That is, suppose that $\mathcal{S}$ is a coalition and that the members of it, and the members of its counter-coalition, agree to $\vec{x}$ as their way of dividing the payoffs. Then, using superadditivity,

$$\sum_{i=1}^{N} x_i = \sum_{P_i \in \mathcal{S}} x_i + \sum_{P_i \in \mathcal{S}^c} x_i = \nu(\mathcal{S}) + \nu(\mathcal{S}^c) \leq \nu(\mathcal{P}).$$

The combination of (6.2) and (6.3) gives the collective rationality condition.

In the game shown in Table 6.1, any 3-tuple $\vec{x}$ which satisfies the conditions

$$x_1 + x_2 + x_3 = 1,$$

and

$$x_1 \geq 1/4, \quad x_2 \geq -1/3, \quad x_3 \geq 0,$$

is an imputation. It is easy to see that there are infinitely many 3-tuples which satisfy these conditions—for example,

$$(1/3, 1/3, 1/3), \quad (1/4, 3/8, 3/8), \quad (1, 0, 0).$$

The following theorem shows that essential games always have many imputations, while inessential ones do not.

THEOREM 6.6. *Let ν be an N-person game in characteristic function form. If ν is inessential, then it has only one imputation, namely,*

$$\vec{x} = (\nu(\{P_1\}), \dots, \nu(\{P_N\})).$$

If ν is essential, then it has infinitely many imputations.

PROOF. Suppose first that ν is inessential, and that $\vec{x}$ is an imputation. If, for some j,

$$x_j > \nu(\{P_j\}),$$

then

$$\sum_{i=1}^{N} x_i > \sum_{i=1}^{N} \nu(\{P_i\}) = \nu(\mathcal{P}).$$

This contradicts collective rationality and shows that, for each i,

$$x_i = \nu(\{P_i\}).$$

Now suppose that ν is essential and let

$$\beta = \nu(\mathcal{P}) - \sum_{i=1}^{N} \nu(\{P_i\}) > 0.$$

For any N-tuple $\vec{\alpha}$ of nonnegative numbers summing to β, we have that

$$x_i = \nu(\{P_i\}) + \alpha_i$$

defines an imputation. Since there are obviously infinitely many choices of $\vec{\alpha}$, there are also infinitely many imputations.   □

For an essential game, there are too many imputations. The problem is to single out those which deserve to be called "solutions." For example, in the game shown in Table 6.1, none of the three imputations listed earlier for it seems likely to occur. For instance, consider the imputation $(1/4, 3/8, 3/8)$. It is unstable because players P_1 and P_2 could form a coalition, gain a total payoff of at least 1, and divide it between them so that each gains more than her entry in $(1/4, 3/8, 3/8)$ gives her.

6.2.1. *Dominance of Imputations*

The following definition attempts to formalize the idea of one imputation being preferred over another by a given coalition.

DEFINITION 6.4. Let ν be a game in characteristic function form, let S be a coalition, and let $\vec{x}$, $\vec{y}$ be imputations. Then we say that $\vec{x}$ *dominates* $\vec{y}$ *through the coalition* S if the following two conditions hold:

- $x_i > y_i$ for all $P_i \in S$.
- $\sum_{P_i \in S} x_i \leq \nu(S)$.

The notation for this relation is

$$\vec{x} \succ_S \vec{y}.$$

The second condition in this definition says that $\vec{x}$ is *feasible*, that is, that the players in S can attain enough payoff so that x_i can actually be paid out to each $P_i \in S$. Since the inequality in the first condition is strict, every player in S does better under $\vec{x}$ (compared to $\vec{y}$).

In the game whose normal form is shown in Table 6.1, we see that $(1/3, 1/3, 1/3)$ dominates $(1, 0, 0)$ through the coalition $\{P_2, P_3\}$, and that $(1/4, 3/8, 3/8)$ dominates $(1/3, 1/3, 1/3)$ through the same coalition. Also, $(1/2, 1/2, 0)$ dominates $(1/3, 1/3, 1/3)$ through $\{P_1, P_2\}$.

6.2.2. *The Core*

It seems intuitively clear that an imputation which is dominated through some coalition would never become permanently established as the way in which the total payoff is actually distributed. Instead, there would be a tendency for the existing coalition to break up and be replaced by one which gives its members a larger share. This idea motivates the following:

DEFINITION 6.5. *Let ν be a game in characteristic function form. The* core *of ν consists of all imputations which are not dominated by any other imputation through any coalition.*

Thus, if an imputation $\vec{x}$ is in the core, there is no group of players which has a reason to form a coalition and replace $\vec{x}$ with a different imputation. The core is the first "solution concept" which we define for N-person co-operative games. As we shall soon see, it has a serious flaw: The core may be empty!

Deciding whether an imputation is in the core seems difficult if we use only the definition. The following theorem makes the job easier.

THEOREM 6.7. *Let ν be a game in characteristic function form with N players, and let $\vec{x}$ be an imputation. Then $\vec{x}$ is in the core of ν if and only if*

$$\sum_{P_i \in S} x_i \geq \nu(S), \tag{6.4}$$

for every coalition S.

PROOF. First, suppose that (6.4) holds for every coalition S. If some other imputation $\vec{z}$ dominates $\vec{x}$ through a coalition S, then

$$\sum_{P_i \in S} z_i > \sum_{P_i \in S} x_i \geq \nu(S),$$

which violates the feasibility condition in Definition 6.4. Thus, $\vec{x}$ is in the core.

Now, suppose that $\vec{x}$ is in the core, and suppose that S is a coalition such that

$$\sum_{P_i \in S} x_i < \nu(S).$$

We must derive a contradiction. We observe that $\mathcal{S} \neq \mathcal{P}$. If this were not true, the collective rationality condition in the definition of imputation would be violated. We next wish to show that there exists $P_j \in \mathcal{S}^c$ such that

$$x_j > \nu(\{P_j\}).$$

If this were not true, we would have, using superadditivity,

$$\sum_{i=1}^{N} x_i < \nu(\mathcal{S}) + \sum_{P_i \in \mathcal{S}^c} x_i \leq \nu(\mathcal{P}),$$

which violates collective rationality. Thus, we can choose $P_j \in \mathcal{S}^c$ such that there exists α with

$$0 < \alpha \leq x_j - \nu(\{P_j\}),$$

and

$$\alpha \leq \nu(\mathcal{S}) - \sum_{P_i \in \mathcal{S}} x_i.$$

Now, with k denoting the number of players in $\mathcal{S}$, we define a new imputation $\vec{z}$ by

$$z_i = x_i + \alpha/k \quad \text{for } P_i \in \mathcal{S},$$
$$z_j = x_j - \alpha,$$

and

$$z_i = x_i \quad \text{for all other } i.$$

Then $\vec{z}$ dominates $\vec{x}$ through $\mathcal{S}$, and so the assumption that $\vec{x}$ is in the core is contradicted. $\square$

The following corollary states a more convenient form of this result. The difference is that we do not have to check separately whether $\vec{x}$ is an imputation.

COROLLARY 6.8. *Let ν be a game in characteristic function form with N players, and let $\vec{x}$ be an N-tuple of numbers. Then $\vec{x}$ is an imputation in the core if and only if the following two conditions hold:*

- $\sum_{i=1}^{N} x_i = \nu(\mathcal{P})$.
- $\sum_{P_i \in \mathcal{S}} x_i \geq \nu(\mathcal{S})$ *for every coalition $\mathcal{S}$.*

PROOF. An imputation in the core certainly satisfies the two conditions. For the other half of the proof, let $\vec{x}$ satisfy both conditions. Applying the second condition to one-player coalitions shows that individual rationality holds. The first condition is collective rationality, and thus $\vec{x}$ is an imputation. It is in the core, by the theorem. $\square$

Let us use this theorem to find the core of the game shown in Table 6.1. By the corollary, a 3-tuple (x_1, x_2, x_3) is an imputation in the core if and only if

$$x_1 + x_2 + x_3 = 1, \tag{6.5}$$
$$x_1 \geq 1/4,$$
$$x_2 \geq -1/3,$$
$$x_3 \geq 0, \tag{6.6}$$
$$x_1 + x_2 \geq 1, \tag{6.7}$$
$$x_1 + x_3 \geq 4/3, \tag{6.8}$$
$$x_2 + x_3 \geq 3/4. \tag{6.9}$$

From (6.5), (6.6), and (6.7), we see that $x_3 = 0$ and $x_1 + x_2 = 1$. From the first of these and (6.8) and (6.9), we have that $x_1 \geq 4/3$ and $x_2 \geq 3/4$. Adding these, we get that $x_1 + x_2 \geq 25/12 > 1$. This is a contradiction, and so we conclude that the core of this game is empty.

As a second example, consider the 3-player game whose characteristic function is given by

$$\nu(\{P_1\}) = -1/2, \tag{6.10}$$
$$\nu(\{P_2\}) = 0,$$
$$\nu(\{P_3\}) = -1/2,$$
$$\nu(\{P_1, P_2\}) = 1/4,$$
$$\nu(\{P_1, P_3\}) = 0,$$
$$\nu(\{P_2, P_3\}) = 1/2,$$
$$\nu(\{P_1, P_2, P_3\}) = 1.$$

The reader should verify that superadditivity holds for this example. We see that a 3-tuple $\vec{x}$ is an imputation in the core of this game if and only if the following all hold:

$$x_1 + x_2 + x_3 = 1,$$
$$x_1 \geq -1/2,$$
$$x_2 \geq 0,$$
$$x_3 \geq -1/2,$$
$$x_1 + x_2 \geq 1/4,$$
$$x_1 + x_3 \geq 0,$$
$$x_2 + x_3 \geq 1/2.$$

This system has many solutions. For example, $(1/3, 1/3, 1/3)$ is in the core.

As a third example, consider the following:

EXAMPLE 6.1 (THE USED CAR GAME). A man named Nixon has an old car he wishes to sell. He no longer drives it, and it is worth nothing to him unless he can sell it. Two people are interested in buying it: Agnew and Mitchell. Agnew values the car at \$500 and Mitchell thinks it is worth \$700. The game consists of each of the prospective buyers bidding on the car, and Nixon either accepting one of the bids (presumably the higher one), or rejecting both of them.

We can write down the characteristic function form of this game directly. Abbreviating the names of the players as N, A, M, we have

$$\nu(\{N\}) = \nu(\{A\}) = \nu(\{M\}) = 0,$$

$$\nu(\{N, A\}) = 500, \quad \nu(\{N, M\}) = 700, \quad \nu(\{A, M\}) = 0,$$

$$\nu(\{N, A, M\}) = 700.$$

The reasoning behind these numbers is as follows. Consider first the one-player coalition $\{N\}$. In the game between this coalition and its counter-coalition, N has only two reasonable pure strategies: (i) Accept the higher bid, or (ii) reject both if the higher bid is less than some lower bound. There exists a joint strategy for the counter-coalition $\{A, M\}$ (namely, both bid zero) such that, if it plays that way, the maximum of N's payoffs over his strategies is zero. By definition of the maximin value, $\nu(\{N\}) = 0$. The characteristic function values of the other two one-player coalitions are zero since the counter-coalition can always reject that player's bid. The coalition $\{N, A\}$ has many joint strategies which result in a payoff to it of \$500, independently of what M does. For example, A could pay N \$500 and take the car. The payoff to N is then \$500, and the payoff to A is zero (the value of the car minus the money). On the other hand, they cannot get more than \$500 without the cooperation of M. Similarly, $\nu(\{N, M\}) = 700$. Finally, the grand coalition has characteristic function value \$700, since that is the largest possible sum of payoffs (attained, for example, if M pays N \$700 for the car).

An imputation (x_N, x_A, x_M) is in the core if and only if

$$x_N, x_A, x_M \geq 0,$$

$$x_N + x_A + x_M = 700,$$

$$x_N + x_A \geq 500, \quad x_N + x_M \geq 700, \quad x_A + x_M \geq 0.$$

These are easily solved to give

$$500 \leq x_N \leq 700, \quad x_M = 700 - x_N, \quad x_A = 0.$$

The interpretation of this solution is that M gets the car with a bid of between \$500 and \$700 (x_N is the bid). Agnew does not get the car, but his presence forces the price up over \$500. This answer is fairly reasonable.

It is consistent with what actually happens in bidding situations, except for one thing. Since the game is cooperative, it is possible for A and M to conspire to bid low. For example, they might agree that Mitchell bids $300, and Agnew bids zero. In exchange, if Nixon accepts the higher bid, Mitchell would pay Agnew $200 for his cooperation. The imputation corresponding to this arrangement is $(300, 200, 200)$. It is not in the core, and so our analysis ignores it.

Another point is worth discussing. Suppose that Agnew and Mitchell play as above, but that Nixon rejects the bid. Thus, he keeps the car (which has no value to him), and the other two neither gain nor lose. The 3-tuple of payoffs is then $(0, 0, 0)$. This is *not* an imputation—individual rationality holds, but collective rationality does not. Referring to our argument concerning collective rationality on page 156, we see that the reason this 3-tuple is ruled out as an imputation is that all three players could do better acting differently. Since the final decision is Nixon's, the conclusion is that he is being irrational in not accepting the bid of $300. Of course, the same reasoning would apply if Mitchell bid $1 instead of $300! The problem here is that, in real life, Nixon might very well reject a bid he thinks is too low. After all, he will have other chances to sell the car, either to Agnew, or to Mitchell, or to someone else.

6.2.3. Constant-Sum Games

The following definition includes, as we shall show, games whose normal forms are zero-sum.

DEFINITION 6.6. Let ν be a game in characteristic function form. We say that ν is *constant-sum* if, for every coalition S, we have

$$\nu(S) + \nu(S^c) = \nu(\mathcal{P}).$$

Further, ν is *zero-sum* if it is constant-sum and if, in addition, $\nu(\mathcal{P}) = 0$.

The game shown in Table 6.1 is constant-sum. The Used Car Game is not, since, for example,

$$\nu(\{N, A\}) + \nu(\{M\}) = 500 + 0 \neq 700 = \nu(\{N, A, M\}).$$

We must be careful here. There is a concept of "zero-sum" both for games in normal form and for games in characteristic function form. They are almost, but not quite, the same. Furthermore, there is a natural definition of *constant-sum* for games in normal form. Again, the two concepts of "constant-sum" are not quite the same. In Exercise (2) of this section the reader is asked to prove that the characteristic function form of a game which is zero-sum in its normal form is zero-sum in its characteristic function form, according to the definition above. We will prove a similar statement for constant-sum games. However, the converse is false—it is possible for a game which is not constant-sum in its normal form to be constant-sum in its characteristic function form. An example of this phenomenon appears in Exercise (4).

We make a formal definition of constant-sum for games in normal form.

DEFINITION 6.7. Let $\vec{\pi}$ be an N-person game in normal form. Then we say that $\vec{\pi}$ is *constant-sum* if there is a constant c such that

$$\sum_{i=1}^{N} \pi_i(x_1, \ldots, x_N) = c,$$

for all choices of strategies $x_1, \ldots, x_N$ for players $P_1, \ldots, P_N$, respectively.

If $c = 0$, this reduces to zero-sum.

THEOREM 6.9. *If an N-person game $\vec{\pi}$ is constant-sum in its normal form, then its characteristic function is also constant-sum.*

PROOF. Let c be the constant value of $\vec{\pi}$ appearing in the previous definition. We define a new game $\vec{\tau}$ by subtracting c/N from every payoff in $\vec{\pi}$. Thus,

$$\tau_i(x_1, \ldots, x_N) = \pi_i(x_1, \ldots, x_N) - c/N,$$

for every choice of i, and for all choices of strategies. Then $\vec{\tau}$ is zero-sum. By Exercise (2), the characteristic function μ of $\vec{\tau}$ is zero-sum. But, it is easy to see that the characteristic function ν of $\vec{\pi}$ is related to μ by the formula

$$\nu(\mathcal{S}) = \mu(\mathcal{S}) + kc/N,$$

where k is the number of players in the coalition $\mathcal{S}$. From this, it is clear that ν is constant-sum. □

The following theorem contains bad news about cores.

THEOREM 6.10. *If ν is both essential and constant-sum, then its core is empty.*

PROOF. Suppose ν has N players:

$$\mathcal{P} = \{P_1, \ldots, P_N\}.$$

The idea of the proof is to assume both that ν is constant-sum and that there is an imputation $\vec{x}$ in its core, and then to prove that ν is inessential. For any player P_j, we have, by individual rationality,

$$x_j \geq \nu(\{P_j\}). \tag{6.11}$$

Since $\vec{x}$ is in the core, we have

$$\sum_{i \neq j} x_i \geq \nu(\{P_j\}^c).$$

Adding these inequalities, we get, using collective rationality,

$$\nu(\mathcal{P}) = \sum_{i=1}^{N} x_i \geq \nu(\{P_j\}) + \nu(\{P_j\}^c) = \nu(\mathcal{P}),$$

by the constant-sum property. It follows that (6.11) is actually an equality. Since it holds for every j, we have

$$\nu(\mathcal{P}) = \sum_{i=1}^{N} \nu(\{P_i\}),$$

which says that ν is inessential. $\square$

6.2.4. *A Voting Game*

The theory of cooperative games has been applied to several problems involving voting. For example, the distribution of power in the United Nations Security Council has been studied in this way (see [Jon80]), as has the effect that the Electoral College method of electing U.S. presidents has had on the relative strengths of voters in different states (see [Owe82]). The game presented here is rather small, but shows some of the features of the lifesize ones.

EXAMPLE 6.2 (LAKE WOBEGON GAME). The municipal government of Lake Wobegon, Minnesota, is run by a City Council and a Mayor. The Council consists of six Aldermen and a Chairman. A bill can become a law in Lake Wobegon in two ways. These are:

- A majority of the Council (with the Chairman voting only in case of a tie among the Aldermen) approves it and the Mayor signs it.
- The Council passes it, the Mayor vetoes it, but at least six of the seven members of the Council then vote to override the veto. (In this situation, the Chairman always votes.)

The game consists of all eight people involved signifying approval or disapproval of the given bill.

In its normal form, the payoffs would be in units of "power" gained by being on the winning side. It is easier to set up the characteristic function form directly than to work with the normal form. Let us call a coalition among the eight players a *winning* coalition if it can pass a bill into law. For example, a coalition consisting of any three Aldermen, the Chairman, and the Mayor is winning. We call a coalition which is not winning a *losing* coalition. Thus, the coalition consisting only of four Aldermen is a losing one (since they do not have the votes to override a mayoral veto). Define $\nu(\mathcal{S}) = 1$ if the coalition $\mathcal{S}$ is winning, and $\nu(\mathcal{S}) = 0$ if $\mathcal{S}$ is losing. Since every one-player coalition is losing, and the grand coalition is winning, an 8-tuple

$$(x_M, x_C, x_1, \ldots, x_6)$$

is an imputation if and only if

$$x_M, x_C, x_1, \ldots, x_6 \geq 0 \text{ and } x_M + x_C + x_1 + \cdots + x_6 = 1.$$

Here, M means Mayor, C means Chairman, and $1, \ldots, 6$ denote the Aldermen. In Exercise (9), the reader is asked to verify that Lake Wobegon is not constant-sum. Nevertheless, its core is empty. We have the following:

THEOREM 6.11. *Lake Wobegon has an empty core.*

PROOF. Suppose, on the contrary, that $(x_M, x_C, x_1, \ldots, x_6)$ is in the core. Now, any coalition consisting of at least six members of the Council is winning. Thus

$$x_C + x_1 + \cdots + x_6 \geq 1,$$

and the same inequality holds if any one of the terms in it is dropped. Since all x's are nonnegative, and the sum of all eight is 1, this implies that all the x's in the inequality above are zero. This is a contradiction. □

Lake Wobegon is an example of an important category of games.

DEFINITION 6.8. A game ν in characteristic function form is called *simple* if all the following hold:

- $\nu(\mathcal{S})$ is either 0 or 1, for every coalition $\mathcal{S}$.
- $\nu(\text{the grand coalition}) = 1$.
- $\nu(\text{any one-player coalition}) = 0$.

In a simple game, a coalition $\mathcal{S}$ with $\nu(\mathcal{S}) = 1$ is called a winning coalition, and one with $\nu(\mathcal{S}) = 0$ is called losing.

Exercises

(1) In the game of The Used Car, suppose Agnew and Mitchell agree that Agnew is to bid $300, and Mitchell zero. Further, they agree that if Nixon accepts the bid, then Agnew will pay Mitchell $200 for his cooperation. If this happens, what is the 3-tuple of payoffs? Is it an imputation? If not, explain what is wrong with it.

(2) Prove that if $\tilde{\pi}$ is a zero-sum game in normal form, then its characteristic function form is also zero-sum.

(3) Referring to Exercise (6) on page 155, for which values of the parameters is the game constant-sum?

(4) Modify the game of Couples [see Exercise (5), page 155] in the following way: If P_1 and P_2 choose each other, and P_3 chooses P_1, then the payoff to P_3 is -2. All other payoffs are the same as before. Prove that the modified game is not zero-sum in its normal form, but is zero-sum in its characteristic function form.

(5) Does the game of Exercise (2) on page 155 have nonempty core?

(6) A four-person game is given in characteristic function form as follows:

$$\nu(\{P_1\}) = -1, \quad \nu(\{P_2\}) = 0, \quad \nu(\{P_3\}) = -1, \quad \nu(\{P_4\}) = 0,$$

$$\nu(\{P_1, P_2\}) = 0, \quad \nu(\{P_1, P_3\}) = -1, \quad \nu(\{P_1, P_4\}) = 1,$$

$$\nu(\{P_2, P_3\}) = 0, \quad \nu(\{P_2, P_4\}) = 1, \quad \nu(\{P_3, P_4\}) = 0,$$

$$\nu(\{P_1, P_2, P_3\}) = 1, \quad \nu(\{P_1, P_2, P_4\}) = 2,$$

$$\nu(\{P_1, P_3, P_4\}) = 0, \quad \nu(\{P_2, P_3, P_4\}) = 1,$$

$$\nu(\{P_1, P_2, P_3, P_4\}) = 2, \quad \nu(\emptyset) = 0.$$

Verify that ν is a characteristic function. Is the core of this game nonempty?

(7) Does the game of Exercise (3), page 155, have nonempty core?

(8) Let ν be a game in characteristic function form. (i) Prove that the set of all imputations is convex. (ii) Prove that the core is convex.

(9) Verify that Lake Wobegon is not a constant-sum game.

6.3. Strategic Equivalence

Consider two games ν and μ in characteristic function form. Suppose that the number of players is the same for both of them. The question we are now concerned with is this: When can we say that ν and μ are "essentially" the same? Suppose, for example, that we merely change the units in which the payoffs are computed. For instance, this would be the result of converting from U.S. dollars to Swiss francs. Such a change would not change the analysis of the game in any way. This change of units is equivalent to multiplying the characteristic function by a positive constant.

Another modification of a game which should have no effect on the mathematical analysis of it is this: Suppose each player P_i receives a fixed amount c_i, independently of how she plays. (Of course, c_i could be negative, in which case, its absolute value represents a fixed payment for the privilege of playing.) Since the players can do nothing to change the c_i's, they would play as if these fixed amounts were not present. Combining the two modifications just discussed leads to the following:

DEFINITION 6.9. Let ν and μ be two games in characteristic function form with the same number N of players. Then μ is *strategically equivalent* to ν if there exist constants $k > 0$, and $c_1, \ldots, c_N$ such that, for every coalition $\mathcal{S}$,

$$\mu(\mathcal{S}) = k\nu(\mathcal{S}) + \sum_{P_i \in \mathcal{S}} c_i. \qquad (6.12)$$

Note first that ν and μ really play symmetric roles in this definition. That is, (6.12) can be solved for ν to give

$$\nu(\mathcal{S}) = (1/k)\mu(\mathcal{S}) + \sum_{P_i \in \mathcal{S}} (-c_i/k),$$

which has the same form as (6.12).

For an example, the game whose normal form appears in Table 6.1 has characteristic function

$$\nu(\mathcal{P}) = 1, \quad \nu(\emptyset) = 0,$$

$$\nu(\{P_1, P_2\}) = 1, \quad \nu(\{P_1, P_3\}) = 4/3, \quad \nu(\{P_2, P_3\}) = 3/4,$$

$$\nu(\{P_1\}) = 1/4, \quad \nu(\{P_2\}) = -1/3, \quad \nu(\{P_3\}) = 0.$$

Letting $k = 2$, and c_1, c_2, c_3 be $-1, 0, -1$, respectively, we have that μ is strategically equivalent to ν, where

$$\mu(\mathcal{P}) = (2)1 + (-1 + 0 - 1) = 0, \quad \mu(\emptyset) = (2)0 = 0,$$

$$\mu(\{P_1, P_2\}) = (2)1 + (-1 + 0) = 1,$$

$$\mu(\{P_1, P_3\}) = (2)(4/3) + (-1 - 1) = 2/3, \quad \mu(\{P_2, P_3\}) = 1/2,$$

$$\mu(\{P_1\}) = -1/2, \quad \mu(\{P_2\}) = -2/3, \quad \mu(\{P_3\}) = -1.$$

In this example, μ is zero-sum.

We emphasize that if ν and μ are strategically equivalent, then the players would play the same in both games. That is, the relative likelihoods of the various coalitions forming would be the same, and the members of the coalitions would adopt the same joint strategies.

We prove the following:

THEOREM 6.12. *If ν and μ are strategically equivalent, and ν is inessential, then so is μ. Thus, if ν is essential, so is μ.*

PROOF. Assuming ν is inessential, compute

$$
\begin{aligned}
\sum_{i=1}^{N} \mu(\{P_i\}) &= \sum_{i=1}^{N} (k\nu(\{P_i\}) + c_i) \\
&= k\sum_{i=1}^{N} \nu(\{P_i\}) + \sum_{i=1}^{N} c_i \\
&= k\nu(\mathcal{P}) + \sum_{i=1}^{N} c_i \\
&= \mu(\mathcal{P}).
\end{aligned}
$$

This shows that μ is inessential. By symmetry, if μ is inessential, so is ν. Thus, if ν is essential, so is μ. □

6.3.1. Equivalence and Imputations

Suppose that ν and μ are strategically equivalent, that is, that the relation given by (6.12) holds. Then the following theorem relates the imputations for ν to those for μ.

THEOREM 6.13. Let ν and μ be strategically equivalent N-person games. Then we have:

- An N-tuple $\vec{x}$ is an imputation for ν if and only if $k\vec{x} + \vec{c}$ is an imputation for μ.
- An imputation $\vec{x}$ dominates an imputation $\vec{y}$ through a coalition S with respect to ν if and only if $k\vec{x} + \vec{c}$ dominates $k\vec{y} + \vec{c}$ with respect to μ through the same coalition.
- An N-tuple $\vec{x}$ is in the core of ν if and only if $k\vec{x} + \vec{c}$ is in the core of μ.

PROOF. Suppose that $\vec{x}$ is an imputation for ν. Then, for $1 \leq i \leq N$,

$$
\mu(\{P_i\}) = k\nu(\{P_i\}) + c_i \leq kx_i + c_i,
$$

which verifies individual rationality since $kx_i + c_i$ is the ith component of $k\vec{x} + \vec{c}$. As for collective rationality,

$$
\begin{aligned}
\mu(\mathcal{P}) &= k\nu(\mathcal{P}) + \sum_{i=1}^{N} c_i \\
&= k\sum_{i=1}^{N} x_i + \sum_{i=1}^{N} c_i \\
&= \sum_{i=1}^{N}(kx_i + c_i).
\end{aligned}
$$

Thus, $k\vec{x} + \vec{c}$ is an imputation for μ. The converse of this statement is true because of the symmetry of ν and μ. The other two statements of the theorem are proved in a way similar to the proof of the first. $\square$

This theorem validates our belief that if we are studying a game in characteristic function form, then we are simultaneously studying *all* games which are strategically equivalent to it. In case ν and μ are strategically equivalent, then we will use phrases like "ν and μ are the same up to strategic equivalence" to emphasize this fact.

6.3.2. *(0,1)-Reduced Form*

Part of the usefulness of the observation we have just made is that we can replace a game by another one whose characteristic function is particularly easy to work with.

DEFINITION 6.10. A characteristic function μ is in $(0,1)$-*reduced form* if both the following hold:

- $\mu(\{P\}) = 0$ for every player P.
- $\mu(\mathcal{P}) = 1$.

A game in $(0,1)$-reduced form is obviously essential. Conversely, it is also true that, up to strategic equivalence, every essential game is in $(0,1)$-reduced form. We have the following:

THEOREM 6.14. *If ν is an essential game, then ν is strategically equivalent to a game μ in $(0,1)$-reduced form.*

PROOF. Define

$$
k = \frac{1}{\nu(\mathcal{P}) - \sum_{i=1}^{N} \nu(\{P_i\})} > 0,
$$

and, for $1 \le i \le N$, define

$$c_i = -k\nu(\{P_i\}).$$

Then μ is defined by (6.12). The easy verification that μ is in $(0,1)$-reduced form is left to the reader. $\square$

A simple game is already in $(0,1)$-reduced form.

Let us do some computation for the game whose normal form is given in Table 6.1. From the formulas in the proof, we have

$$k = \frac{1}{1-(-1/12)} = 12/13,$$

and

$$c_1 = -(12/13)(1/4) = -3/13, \quad c_2 = 4/13, \quad c_3 = 0.$$

Then μ is given by

$$\mu(\mathcal{P}) = 1, \quad \mu(\emptyset) = 0,$$
$$\mu(\text{any one-player coalition}) = 0,$$
$$\mu(\{P_1, P_2\}) = (12/13)(1) - 3/13 + 4/13 = 1,$$

and

$$\mu(\{P_1, P_3\}) = \mu(\{P_2, P_3\}) = 1.$$

For this game μ, we see three things immediately:

- All three two-person coalitions are equally good.
- If a two-person coalition forms, the players will probably divide the payoff equally (since the players have completely symmetric roles).
- There is no advantage to a two-player coalition in bringing in the third player to form the grand coalition.

We conclude that one of the two-player coalitions will form, the players in it will split the payoff, and the third player will be left out in the cold. Thus, the prevailing imputation will be either $(1/2, 1/2, 0)$, $(1/2, 0, 1/2)$, or $(0, 1/2, 1/2)$. Our analysis is unable to predict which of the three two-player coalitions will actually form.

If we transform these conclusions back into terms of ν, we see that one of the two-player coalitions will form. The prevailing imputation will be one of three possibilities which can be computed using the relationship between μ and ν. This computation will be carried out shortly.

The reader may verify that the $(0,1)$-reduced form μ of the game given in (6.10) on page 161 is such that

$$\mu(\{P_1, P_2\}) = 3/8, \quad \mu(\{P_1, P_3\}) = \mu(\{P_2, P_3\}) = 1/2.$$

In this example, the nice symmetry of the previous game is lost. We can safely say that the two-player coalitions seem weak, and that the grand coalition is likely to form. Stated differently, any of the two-player coalitions would profit by recruiting the third player. To make a guess about what the final imputation might be is hazardous. From the earlier discussion of the game, we know that the core is large.

The $(0, 1)$-reduced form of a game is unique. This fact follows from the following theorem and from Exercise (1).

THEOREM 6.15. *Suppose μ and ν are N-person games in $(0, 1)$-reduced form. If they are strategically equivalent, then they are equal.*

PROOF. By definition of strategic equivalence, there exist constants $k > 0$, and $c_1, \ldots, c_N$, such that

$$\mu(\mathcal{S}) = k\nu(\mathcal{S}) + \sum_{P_i \in \mathcal{S}} c_i,$$

for every coalition $\mathcal{S}$. To prove that μ and ν are equal, we must show that $k = 1$ and $c_i = 0$ for all i. Since both characteristic functions are zero for all one-player coalitions, we see that $c_i = 0$ for all i. Since both characteristic functions are 1 for the grand coalition, we see that $k = 1$. □

6.3.3. *Classification of Small Games*

Up to strategic equivalence, the number of games with two or three players is limited, as the following three theorems show. All three are easy to prove.

THEOREM 6.16. *A two-player game in characteristic function form is either inessential or strategically equivalent to ν, where*

$$\nu(\text{the grand coalition}) = 1, \quad \nu(\emptyset) = 0,$$

$$\nu(\text{either one-player coalition}) = 0.$$

In the case of constant-sum games with three players, we have the following:

THEOREM 6.17. *Every three-player constant-sum game in characteristic function form is either inessential or is strategically equivalent to ν, where*

$$\nu(\text{the grand coalition}) = 1, \quad \nu(\emptyset) = 0,$$

$$\nu(\text{any two-player coalition}) = 1,$$

$$\nu(\text{any one-player coalition}) = 0.$$

Since the phrase "three-player, constant-sum, essential game in $(0,1)$-reduced form" is fairly clumsy, we will abbreviate it by speaking of the game THREE. The previous theorem says that every essential constant-sum game with three players is strategically equivalent to THREE. In particular, the game whose normal form is given in Table 6.1 is strategically equivalent to THREE.

Up to strategic equivalence, the three-player games which are not necessarily constant-sum form a three-parameter family. We have the following:

THEOREM 6.18. *Every three-player game in characteristic function form is either inessential or there exist constants a, b, c satisfying*

$$0 \le a \le 1, \quad 0 \le b \le 1, \quad 0 \le c \le 1,$$

such that the game is strategically equivalent to ν, where

$$\nu(\text{the grand coalition}) = 1, \quad \nu(\emptyset) = 0.$$

and

$$\nu(\text{any one-player coalition}) = 0,$$

$$\nu(\{P_1, P_2\}) = a, \quad \nu(\{P_1, P_3\} = b, \quad \nu(\{P_2, P_3\}) = c.$$

Exercises

(1) Prove that if λ is strategically equivalent to μ, and if μ is strategically equivalent to ν, then λ is strategically equivalent to ν.

(2) Suppose that ν and μ are strategically equivalent. Prove that if one of them is constant-sum, then so is the other.

(3) Prove that a constant-sum game is strategically equivalent to a zero-sum game.

(4) The Zero Game is such that its characteristic function is equal to zero for *every* coalition. (It's not much fun to play.) Prove that every inessential game is strategically equivalent to the Zero Game.

(5) Compute the $(0,1)$-reduced form for the Used Car Game.

(6) Compute the $(0,1)$-reduced form of the game in Exercise (6) on page 167.

(7) Compute the $(0,1)$-reduced form of the game referred to in Exercise (3), page 155.

(8) Compute the $(0,1)$-reduced form of Couples.

(9) State and prove a theorem which characterizes four-player constant-sum games.

6.4. Two Solution Concepts

As a solution concept for the games we are studying, the core is flawed. Often, there are no imputations in the core; when there are, there are often so many that we have no reasonable way to decide which ones are actually likely to occur. Several attempts have been made over the years to define more acceptable solution concepts. We discuss two of them here.

6.4.1. *Stable Sets of Imputations*

The definition of a stable set of imputations is fairly natural:

DEFINITION 6.11. Let X be a set of imputations for a game in characteristic function form. Then we say that X is *stable* if the following two conditions hold:

- (*Internal Stability*) No imputation in X dominates any other imputation in X through any coalition.
- (*External Stability*) If $\bar{y}$ is any imputation outside X, then it is dominated through some coalition by some imputation inside X.

This idea was introduced by von Neumann and Morgenstern in [vNM44]. They went so far as to call a stable set a *solution* of the game. Before discussing their reasons for doing so, let us note that an imputation inside a stable set may be dominated by some imputation outside. Of course, that outside imputation is, in turn, dominated by some other imputation inside (by external stability). This seems wrong at first because we tend to assume that the relation of dominance through a coalition is "transitive," but this is not true. Now, let us consider what might happen in a game with a stable set X. By external stability, an imputation outside X seems unlikely to become established. There exists a coalition (possibly many) which definitely prefers one of the imputations inside X. Therefore, there would be a tendency toward a shift to such an inside imputation. By internal stability, all the imputations inside X are equal as far as domination through coalitions is concerned. Presumably, the one which actually prevails would be chosen in some way not amenable to mathematical analysis—pure chance, custom, precedent, etc. But, now, there is a problem. We just mentioned that there may be an imputation outside X which dominates a given imputation inside. Why would not a new coalition form in order to take advantage of this outside imputation? If this happens, X is abandoned (until another coalition forms and the game moves back into X). Another way of looking at this problem is that, in fact, the stable set X is not unique. There may well be others (as later examples will show). Why should the game stay inside any given one of them? Of course, it could be that the history of the game might actually turn out to be a chaotic series of formations and dissolutions of coalitions, of moving into and out of various stable sets. After all, real life often looks that way. But, if that is the way the game goes, why call a stable set a solution?

The explanation given by von Neumann and Morgenstern for calling a stable set a solution is that stable sets correspond to "sound" or "accepted" *standards of behavior*. That is, they represent the principles of conduct which the community (that is, the set of players) accepts as ethically right. Certain imputations (ones outside a certain stable set) are condemned as wrong, and thus can never be established, even though coalitions exist which would benefit from them. For example, in civilized societies, the poorest people are not actually left to die on the street, even though richer people have to pay more taxes in order to prevent this happening.

The von Neumann-Morgenstern concept of a stable set as a solution is thus based on extra-mathematical grounds. One can accept it as reasonable or unreasonable (depending, perhaps, on the game), but there is room for honest disagreement.

Let us consider, for example, the game THREE. It has a nice stable set which we have already mentioned. Let

$$X = \{(0, 1/2, 1/2), (1/2, 0, 1/2), (1/2, 1/2, 0)\}.$$

These are the three imputations which we saw as likely to occur for THREE. We prove the following:

THEOREM 6.19. *The set X defined above is a stable set for THREE.*

PROOF. We denote the characteristic function for THREE by μ. First, we verify internal stability. By symmetry, it is enough to show that the imputation $(0, 1/2, 1/2)$ does not dominate $(1/2, 0, 1/2)$ through any coalition. Now, we see that the only possible coalition through which this domination could occur is $\{P_2\}$. But

$$\mu(\{P_2\}) = 0 < 1/2.$$

and this violates the feasibility condition in the definition of dominance. Thus, internal stability is proved.

To prove external stability, let $\vec{y}$ be an imputation for THREE outside X. We must show that one of the members of X dominates it through some coalition. Now note that there are at least two values of i for which $y_i < 1/2$. If this were not true, we would have $y_i \geq 1/2$ for two values of i. But, since each y_i is nonnegative, and the y_i's sum to one, this implies that $\vec{y}$ is one of the imputations in X. This contradicts the assumption that $\vec{y}$ is not in X. By symmetry, we may assume that y_1 and y_2 are both less than $1/2$. But, then, $(1/2, 1/2, 0)$ dominates $\vec{y}$ through the coalition $\{P_1, P_2\}$. □

Note that there are imputations outside X which dominate members of X. For example, $(2/3, 1/3, 0)$ dominates $(1/2, 0, 1/2)$ through $\{P_1, P_2\}$. On the other hand, $(0, 1/2, 1/2)$ (a member of X) dominates $(2/3, 1/3, 0)$ through $\{P_2, P_3\}$.

Since every essential constant-sum game with three players is strategically equivalent to THREE, we can use X to obtain a stable set for any such game. For example, let ν be the game whose normal form is shown in Table 6.1. Then

$$\mu(\mathcal{S}) = k\nu(\mathcal{S}) + \sum_{P_i \in \mathcal{S}} c_i,$$

for every coalition $\mathcal{S}$, where $k = 12/13$, $c_1 = -3/13$, $c_2 = 4/13$, and $c_3 = 0$. (See the computation on page 171.) Thus

$$\nu(\mathcal{S}) = (1/k)\mu(\mathcal{S}) + \sum_{P_i \in \mathcal{S}} (-c_i/k).$$

Replacing each imputation $\vec{x}$ in X by $(1/k)\vec{x} - (1/k)\vec{c}$ then gives us a stable set for ν, namely,

$$\{(19/24, 5/24, 0), (19/24, -1/3, 13/24), (1/4, 5/24, 13/24)\}.$$

THREE has other stable sets. We have the following:

THEOREM 6.20. *Let c be any constant such that*

$$0 \le c < 1/2.$$

Then the set of imputations

$$Z_c = \{(c, x_2, x_3) : x_2, x_3 \ge 0, x_2 + x_3 = 1 - c\}$$

is a stable set for THREE.

PROOF. To verify internal stability, let us suppose that (c, x_2, x_3) and (c, x_2', x_3') are both in Z_c, and that the first of these dominates the other through some coalition. Then this coalition is clearly either $\{P_2\}$ or $\{P_3\}$. But this is not possible because the feasibility condition in the definition of domination would be violated.

To prove external stability, let (y_1, y_2, y_3) be an imputation outside Z_c. If $y_1 = c$, then $\vec{y} \in Z_c$. Therefore, there are two cases:

- Suppose $y_1 < c$. Since the y_i's are nonnegative and sum to one, either $y_2 \le 1/2$ or $y_3 \le 1/2$. The two cases are practically the same, so let us assume that $y_2 \le 1/2$. Then choose a positive number α so small that

$$c + \alpha \le 1/2.$$

This is possible because $c < 1/2$. Then the 3-tuple

$$(c, 1/2 + \alpha, 1/2 - c - \alpha) \in Z_c,$$

and dominates (y_1, y_2, y_3) through the coalition $\{P_1, P_2\}$.

- Suppose $y_1 > c$. Then $y_2 + y_3 < 1 - c$. Let

$$\beta = \frac{1 - c - y_2 - y_3}{2} > 0,$$

and let

$$x_2 = y_2 + \beta, x_3 = y_3 + \beta.$$

Then (c, x_2, x_3) is in Z_c, and it dominates (y_1, y_2, y_3) through the coalition $\{P_2, P_3\}$. $\square$

The stable set Z_c is said to be *discriminatory* toward P_1. The idea is that P_2 and P_3 agree to give P_1 the amount c, and to negotiate between themselves about the division of the rest. By symmetry, there are similar stable sets which are discriminatory toward P_2 and P_3, respectively. They are

$$\{(x_1, c, x_3) : x_1, x_3 \geq 0, x_1 + x_3 = 1 - c\},$$

and

$$\{(x_1, x_2, c) : x_1, x_2 \geq 0, x_1 + x_2 = 1 - c\},$$

both defined for $0 \leq c < 1/2$.

For simple games, there are some interesting stable sets of imputations. Let ν be such a game, and call a winning coalition S *minimal* if every coalition properly contained in S is losing. For example, in the Lake Wobegon Game, the coalition consisting only of the six Aldermen is a minimal winning coalition.

THEOREM 6.21. *Let ν be an N-player simple game, and let S be a minimal winning coalition. Then the set X of all imputations $\vec{x}$ such that*

$$x_i = 0 \text{ for } P_i \notin S,$$

is a stable set.

PROOF. To prove internal stability, suppose that $\vec{x}$ and $\vec{z}$ are both in X, and that $\vec{x}$ dominates $\vec{z}$ through some coalition T. Then, clearly, T is properly contained in S. But

$$\sum_{P_i \in T} x_i > 0 = \nu(T),$$

which contradicts the feasibility condition in the definition of domination. Thus, internal stability holds.

To prove external stability, let $\vec{y}$ be an imputation outside X. Define

$$\alpha = 1 - \sum_{P_i \in \mathcal{S}} y_i > 0.$$

Then define, for $P_i \in \mathcal{S}$,

$$x_i = y_i + \alpha/k,$$

where k is the number of players in $\mathcal{S}$; finally, define, for $P_i \notin \mathcal{S}$,

$$x_i = 0.$$

Then $\vec{x}$ is in X and dominates $\vec{y}$ through $\mathcal{S}$. □

In the Lake Wobegon Game, this type of stable set represents a situation in which a winning coalition refuses to share power with those who are not members.

For some years it was hoped, but not proved, that every game had a stable set. However, in 1967 (see [Luc68]), an example of a 10-player game without a stable set was discovered.

There is a discussion of an application of the theory of stable sets to economics (namely, the Edgeworth trading model) in [Jon80].

6.4.2. *Shapley Values*

The second solution concept discussed here is that of the *Shapley value*. The idea was introduced in [Sha53], and is an interesting attempt to define, in a fair way, an imputation which embodies what the players' final payoffs "should" be. It attempts to take into account a player's contribution to the success of the coalitions she belongs to. If the characteristic function of the game is ν, and if $\mathcal{S}$ is a coalition to which player P_i belongs, then the number

$$\delta(P_i, \mathcal{S}) = \nu(\mathcal{S}) - \nu(\mathcal{S} - \{P_i\})$$

is a measure of the amount that P_i has contributed to $\mathcal{S}$ by joining it. These numbers will be used in defining the Shapley value, but, in themselves, are not very revealing. For example, consider the game THREE. For the grand coalition, we have

$$\delta(P_i, \mathcal{P}) = 0,$$

for each player. That is, no one contributes anything. If $\mathcal{S}$ is a two-player coalition, then

$$\delta(P_i, \mathcal{S}) = 1,$$

for each player in $\mathcal{S}$. That is, the sum of the contributions is greater than $\nu(\mathcal{S})$. Both of these results are amusing, but unhelpful.

To begin our derivation of the Shapley value (called ϕ_i, for player P_i), notice that once the players have collectively agreed on an imputation, it might as well be assumed that it is the grand coalition which forms. This is because the condition of collective rationality ensures that the total of all payments (via the imputation) is $\nu(\mathcal{P})$. In any case, we make this assumption, and concentrate on the process by which the grand coalition comes into being. Our assumption is that this process starts with one player; she is joined by a second, and they are later joined by a third, etc. Thus, the process is characterized by an ordered list of the players, with the kth player in the list being the kth one to join. Let us consider a four-person example. Suppose that its characteristic function is given by

$$\nu(\mathcal{P}) = 100, \quad \nu(\emptyset) = 0, \qquad (6.13)$$

$$\nu(\{P_1\}) = 0, \quad \nu(\{P_2\}) = -10, \quad \nu(\{P_3\}) = 10, \quad \nu(\{P_4\}) = 0,$$
$$\nu(\{P_1, P_2\}) = 25, \quad \nu(\{P_1, P_3\}) = 30, \quad \nu(\{P_1, P_4\}) = 10,$$
$$\nu(\{P_2, P_3\}) = 10, \quad \nu(\{P_2, P_4\}) = 10, \quad \nu(\{P_3, P_4\}) = 30,$$
$$\nu(\{P_1, P_2, P_3\}) = 50, \quad \nu(\{P_1, P_2, P_4\}) = 30,$$
$$\nu(\{P_1, P_3, P_4\}) = 50, \quad \nu(\{P_2, P_3, P_4\}) = 40.$$

Here is one ordering of the players through which the grand coalition could form:

$$P_3, P_2, P_1, P_4. \qquad (6.14)$$

There are several other possible orderings. In fact, there are

$$4 \times 3 \times 2 = 4!$$

of them. In general, the number would be $N!$. We think of the choice of the actual ordering by which the grand coalition comes about as a random event. Since there are, in general, $N!$ possibilities, it is reasonable to assign probability $1/N!$ to each of them. Now, given that the grand coalition forms according to the ordering shown above in (6.14), the number

$$\delta(P_1, \{P_3, P_2, P_1\}) = \nu(\{P_3, P_2, P_1\}) - \nu(\{P_3, P_2\}) = 50 - 10 = 40$$

is a measure of the contribution P_1 makes as she enters the growing coalition. In general, the definition of ϕ_i is this: Make the same sort of calculation for each of the $N!$ possible orderings of the players; weight each one by the probability $1/N!$ of that ordering occurring; and add the results.

We will use this definition in two ways. First, we will develop a formula which makes the computation of ϕ_i somewhat easier, and, second, we will prove that $\vec{\phi} = (\phi_1, \ldots, \phi_N)$ is an imputation. Thus, each player will be able to get back exactly the amount contributed. To derive a formula for ϕ_i, note that, among the $N!$ terms in the sum which defines ϕ_i, there are many duplications. Indeed, suppose that we have an ordering of the players such that P_i occurs at position k. Denote by S the set of k players up to and including P_i in this ordering. Then if we permute the part of the ordering coming before P_i, and permute the part coming after it, we obtain a new ordering in which P_i again is in the kth position. Moreover, for both the original and the permuted orderings, the term in the sum defining ϕ_i is

$$\delta(P_i, \mathcal{S}) = \nu(\mathcal{S}) - \nu(\mathcal{S} - \{P_i\}).$$

There are $(k-1)!$ permutations of the players coming before P_i, and $(N-k)!$ permutations of the players coming after P_i. Thus, the term $\delta(P_i, \mathcal{S})$ occurs

$$(N - k)!(k - 1)!$$

times. Letting $|\mathcal{S}|$ denote the number of players in $\mathcal{S}$, we finally get

$$\phi_i = \sum_{P_i \in \mathcal{S}} \frac{(N - |\mathcal{S}|)!(|\mathcal{S}| - 1)!}{N!} \delta(P_i, \mathcal{S}). \tag{6.15}$$

The number ϕ_i is called the *Shapley value* for P_i, and $\vec{\phi}$ is called the *Shapley vector* for the game. Before getting into examples, we mention that there is an alternative way of obtaining this formula. It can be done by listing three axioms which we would want the Shapley value to satisfy. Then, a theorem can be proved to the effect that (6.15) gives the unique value which satisfies the axioms. See [Sha53] or [Vor77] for more information. After considering some examples, we prove that $\vec{\phi}$ is an imputation.

Consider, first, the game whose normal form is shown in Table 6.1. Its characteristic function is given on page 168. To compute ϕ_1, note that there are four coalitions containing P_1, namely,

$$\{P_1\}, \{P_1, P_2\}, \{P_1, P_3\}, \{P_1, P_2, P_3\}.$$

Therefore, (6.15) has four terms in this case. For each of the four coalitions containing P_1, we compute

$$\delta(P_1, \{P_1\}) = 1/4 - 0 = 1/4, \quad \delta(P_1, \{P_1, P_2\}) = 1 - (-1/3) = 4/3,$$

$$\delta(P_1, \{P_1, P_3\}) = 4/3 - 0 = 4/3, \quad \delta(P_1, \{P_1, P_2, P_3\}) = 1 - 3/4 = 1/4.$$

Then

$$\phi_1 = \frac{2!0!}{3!}(1/4) + \frac{1!1!}{3!}(4/3) + \frac{1!1!}{3!}(4/3) + \frac{0!2!}{3!}(1/4) = 11/18.$$

By similar calculations, we have

$$\phi_2 = 1/36, \qquad \phi_3 = 13/36.$$

Notice that $\vec{\phi}$ is an imputation. The Shapley values can be interpreted as reflecting the bargaining power of the players. In this example, P_1's Shapley value is the largest of the three, indicating that he is the strongest; on the other hand, P_2's value is very small. Player P_3 is in the middle. A glance at the characteristic function supports this interpretation.

The Shapley vector for this game can be computed in another way. Note, first, that in the game of THREE, the Shapley vector is surely $(1/3, 1/3, 1/3)$. This is because the players have absolutely symmetric roles in the game. (It is also easy to verify by computation that this is true.) In any case, any three-player constant-sum essential game ν is strategically equivalent to THREE. We can use the transformation introduced in Theorem 6.13 to obtain the Shapley vector for ν.

For the game given in (6.10) on page 161, the reader may verify that the Shapley vector is

$$(1/8, 5/8, 1/4).$$

These numbers seem to reasonably reflect the advantage that player P_2 has in the game. For more discussion of this game, see page 171.

For the Used Car Game, the Shapley values are

$$\phi_N = 433.33\ldots, \quad \phi_A = 83.33\ldots, \quad \phi_M = 183.33\ldots.$$

Thus, Mitchell gets the car for \$433.33, but has to pay Agnew \$83.33 as a bribe for not bidding against him. The Shapley vector indicates that Nixon is in the most powerful position.

THEOREM 6.22. *Let ν be a game in characteristic function form. Then the Shapley vector for ν is an imputation.*

PROOF. To prove individual rationality, we must show that

$$\phi_i \geq \nu(\{P_i\}).$$

Now, by superadditivity, if $P_i \in \mathcal{S}$,

$$\delta(P_i, \mathcal{S}) = \nu(\mathcal{S}) - \nu(\mathcal{S} - \{P_i\}) \geq \nu(\{P_i\}).$$

Thus,

$$\phi_i \geq \left(\sum_{P_i \in \mathcal{S}} \frac{(N - |\mathcal{S}|)!(|\mathcal{S}| - 1)!}{N!} \right) \nu(\{P_i\}).$$

The sum in this inequality is the sum of the probabilities of the different orderings of the players. As such, it must equal 1, and so

$$\phi_i \geq \nu(\{P_i\}).$$

To prove collective rationality, consider

$$\sum_{i=1}^{N} \phi_i = \sum_{i=1}^{N} \sum_{P_i \in \mathcal{S}} \frac{(N - |\mathcal{S}|)!(|\mathcal{S}| - 1)!}{N!} \delta(P_i, \mathcal{S}).$$

In this double sum, let us fix our attention on the terms involving $\nu(T)$, where T is a fixed nonempty coalition which is not equal to $\mathcal{P}$. Then there are two kinds of terms involving $\nu(T)$—those with a positive coefficient (when $T = S$):

$$\frac{(N - |T|)!(|T| - 1)!}{N!},$$

and those with a negative coefficient (when $T = S - \{P_i\}$):

$$-\frac{(N - 1 - |T|)!|T|!}{N!}.$$

The first kind occurs $|T|$ times (one for each member of T), and the second kind occurs $N - |T|$ times (once for each player outside T). Thus the coefficient of $\nu(T)$ in the double sum is

$$\frac{|T|(N - |T|)!(|T| - 1)!}{N!} - \frac{(N - |T|)(N - 1 - |T|)!|T|!}{N!} =$$
$$\frac{(N - |T|)!|T|!}{N!} - \frac{(N - |T|)!|T|!}{N!} = 0.$$

Therefore, the only terms left in the double sum are those involving the grand coalition, and those involving the empty coalition. We have, since $\nu(\emptyset) = 0$,

$$\sum_{i=1}^{N} \phi_i = \frac{N(0!)(N - 1)!}{N!}\nu(\mathcal{P}) = \nu(\mathcal{P}).$$

This proves collective rationality. $\square$

For another example, we compute the Shapley values for the Lake Wobegon Game. We start with ϕ_M. The nonzero terms in (6.15) are those for which $S - \{M\}$ is a losing coalition, but S is winning. That is, they are coalitions which, if the Mayor is removed, can pass a bill but cannot override a mayoral veto. A little thought shows that there are four types of these winning coalitions, namely,

(1) S contains the Mayor, three of the Aldermen, and the Chairman.
(2) S contains the Mayor and four Aldermen.
(3) S contains the Mayor, four Aldermen, and the Chairman.
(4) S contains the Mayor and five Aldermen.

There are

$$\binom{6}{3} = 20$$

sets of the first type. Since $|S| = 5$,

$$\frac{(N - |S|)!(|S| - 1)!}{N!} = \frac{(8 - 5)!(5 - 1)!}{8!} = 1/280.$$

Thus, the contribution to ϕ_M from these sets is

$$20/280 = 1/14.$$

There are 15 sets of the second type, and the contribution to ϕ_M from them is

$$(15)\frac{(8-5)!(5-1)!}{8!} = 3/56.$$

There are 15 sets of the third type, and the contribution to ϕ_M from them is

$$(15)\frac{(8-6)!(6-1)!}{8!} = 5/56.$$

Finally, there are 6 sets of the fourth type, and the contribution to ϕ_M from them is

$$(6)\frac{(8-6)!(6-1)!}{8!} = 1/28.$$

Adding these four numbers, we get

$$\phi_M = 1/14 + 3/56 + 5/56 + 1/28 = 1/4.$$

To compute ϕ_C, note that there are only two types of sets which make a contribution to ϕ_C, namely,

(1) S contains the Chairman, three Aldermen, and the Mayor. (In this case, the vote among the Aldermen is a tie, the Chairman votes to approve, and the Mayor signs it.)

(2) S contains the Chairman and five Aldermen. (In this case, the bill is vetoed, but with the Chairman's vote, the veto is overridden.)

There are 20 sets of the first type, and 6 of the second. Then

$$\phi_C = (20)\frac{(8-5)!(5-1)!}{8!} + (6)\frac{(8-6)!(6-1)!}{8!} = 3/28.$$

Now, the sum of the ϕ's is 1, and all the ϕ_i's are surely equal (by symmetry). Thus, for each i,

$$\phi_i = (1/6)(1 - 1/4 - 3/28) = 3/28.$$

These results say that the Mayor has much more power than an Alderman or the Chairman. It turns out that the Chairman's power is exactly equal to that of an Alderman. [Also, see Exercise (9).]

Exercises

(1) Prove that the set

$$\{(x, 0, 700 - x) : 0 \leq x \leq 700\}$$

is a stable set of imputations for The Used Car Game.

(2) Let X be a stable set of imputations for a game ν. Prove that the core is a subset of X.

(3) List all the minimal winning coalitions for the Lake Wobegon Game.

(4) Compute the Shapley vector for the game in Exercise (6) on page 167.

(5) Compute the Shapley vector for the game referred to in Exercise (3) on page 155.

(6) Verify the Shapley values given in the text for the Used Car Game.

(7) Compute the Shapley value for each player in the game given in (6.13) on page 179.

(8) What is the Shapley vector for an inessential game?

(9) Consider a committee with $2m$ ordinary members and a Chairman who only votes in case of a tie. Prove that the power of the Chairman (defined as the Shapley value) is equal to that of an ordinary member.

(10) The town of Bedford Falls has a bicameral Council. The Upper House has three members, and the Lower House has seven. To pass a bill, a majority is needed in each House (there is no mayoral veto). Compute the power of each member of each House (defined as the Shapley value).

(11) Prove that if a game has a stable set containing only one imputation, then it is inessential.

7
Game-Playing Programs

Computer programs which are capable of playing chess and other games at a high level of skill have become familiar to most of us in recent years. In fact, there are chess-playing programs which can beat all but the best human players, and many experts think that the world champion will be a program within the next few years (see [Lev84]). Our aim in this chapter is to discuss the mathematical background involved in these programs.

The actual writing of such a program involves more than this mathematical background, of course. There is a part of the process which requires an intimate knowledge and "feel" for the game. The writing of a game-playing program is an interesting exercise in both game theory and computer programming. The moment when the programmer realizes that his or her creation plays the game better than its creator is both exhilarating and strangely unsettling. The famous Baron von Frankenstein must have had similar feelings.

The extensive form of a game is used in this chapter. The first task is to discuss algorithms for computing a player's optimal move. Throughout this chapter, the games studied are two-person zero-sum games of perfect information. Thus each player has an optimal pure strategy (by Theorem 1.9). In addition, we assume that there are no chance moves. This last assumption could be removed but we make it so as to simplify the discussion.

7.1. Three Algorithms

Let T be the tree for a game satisfying the conditions mentioned above. Let A and B be the two players. We assume that the players move alternately until a terminal vertex is reached. The algorithms of this section compute optimal first moves for the player who owns the root. In other words, the move produced by the algorithms is an edge from the root which belongs to an optimal pure strategy for that player.

7.1.1. *The Naive Algorithm*

The following algorithm is simple enough to show the general idea, but is not as efficient as the other two algorithms of this section. The quantities $+\infty$ and $-\infty$ which appear in the algorithm would, in a program, be replaced by numbers which are greater than and less than, respectively, any possible payoff.

ALGORITHM 7.1 (NAIVE). *We are given a game tree T as just described. An optimal move from the root for the player owning it is computed.*

(1) *We define two functions $\pi(u)$ and $\sigma(u)$; $\pi(u)$ is defined on the set of all vertices and is real-valued; $\sigma(u)$ is defined on the set of all nonterminal vertices, and takes its values in the set of vertices. They are defined recursively as follows.*

   (a) *If u is terminal, let*

$$\pi(u) = p_A(u),$$

   where $p_A(u)$ is the payoff to player A at terminal vertex u.

   (b) *If u belongs to A, let $\{v_1, \ldots, v_d\}$ be the children of u and compute as follows.*

   (i) *Set $i = 1$, $s = -\infty$, $w = v_1$.*
   (ii) *Set $t = \pi(v_i)$.*
   (iii) *If $t > s$, set $s = t$ and $w = v_i$.*
   (iv) *Set $i = i + 1$.*
   (v) *If $i \leq d$ go to (ii).*
   (vi) *Set $\pi(u) = s$ and $\sigma(u) = w$.*

   (c) *If u belongs to B, let $\{v_1, \ldots, v_d\}$ be the children of u and compute as follows.*

   (i) *Set $i = 1$, $s = +\infty$, $w = v_1$.*
   (ii) *Set $t = \pi(v_i)$.*
   (iii) *If $t < s$, set $s = t$ and $w = v_i$.*
   (iv) *Set $i = i + 1$.*
   (v) *If $i \leq d$ go to (ii).*
   (vi) *Set $\pi(u) = s$ and $\sigma(u) = w$.*

(2) *An optimal move for the player owning the root r is given by $\sigma(r)$.*

The computation in case u belongs to A results in finding the child v of u for which $\pi(v)$ is a maximum. Then $\sigma(u)$ is that child, and $\pi(u)$ is the value of that maximum. In case u belongs to B, the computation is the same except that *maximum* is replaced by *minimum*. It is intuitively plausible that the restriction of the function $\sigma(u)$ to the vertices owned by A is an optimal choice function for A, that the restriction of $\sigma(u)$ to the vertices owned by B is an optimal choice function for B, and that $\pi(r)$ is the value of the game. Of course, the corresponding statements should hold if T is replaced by any of its cuttings. The formal proof of these facts is omitted; it is similar to the proof of Theorem 1.9. The theorem here is as follows:

THEOREM 7.1. *Let T be a game tree as described above. Apply the naive algorithm to compute the functions $\sigma(u)$ and $\pi(u)$. Then, if v is a vertex of T, we have:*

- $\pi(v)$ *is the value of the game whose tree is the cutting T_v.*
- *The restriction of $\sigma(u)$ to the vertices owned by player A in the cutting T_v gives an optimal strategy for A in the game T_v.*
- *The restriction of $\sigma(u)$ to the vertices owned by player B in the cutting T_v gives an optimal strategy for B in the game T_v.*

We now apply the algorithm to an example. The tree is shown in Figure 7.1. The root is in the center and the circled numbers labeling the terminal vertices are the payoffs to player A. The other numbers at the vertices are intended only to identify them. We assume that the algorithm scans children in increasing numerical order.

Let us trace the computation of $\sigma(r)$. In this computation, the algorithm first calls for the computation of $\pi(1)$. In order to compute $\pi(1)$, $\pi(4)$ is needed. The children of vertex 4 are all terminal and vertex 4 belongs to A, and so we have that $\pi(4) = 5$ and $\sigma(4) = 10$. Similarly, $\pi(5) = 7$ and $\sigma(5) = 13$. Thus, since vertex 1 belongs to B, $\pi(1) = 5$ and $\sigma(1) = 4$. Then the algorithm calls for $\pi(2)$. To compute this, $\pi(6)$ is needed. We see that $\pi(6) = 4$ and $\sigma(6) = 17$. Then, $\pi(7) = 8$ and $\sigma(7) = 21$. Thus, $\pi(2) = 4$ and $\sigma(2) = 6$. Finally, $\pi(3) = 2$ and $\sigma(3) = 9$. We get that $\sigma(r) = 1$ and $\pi(r) = 5$.

7.1.2. *The Branch and Bound Algorithm*

Let us examine further the calculation just carried out for the game of Figure 7.1. In the calculation of $\pi(2)$, the value $\pi(6) = 4$ was computed first. At this point, we (but not the naive algorithm) would know that $\pi(2) \leq 4$. Since it is already known that $\pi(1) = 5$, it is clear that $\sigma(r) \neq 2$. This means that the calculation of $\pi(7)$ is unnecessary. Stated differently, the payoffs labeling vertices 19, 20, and 21 could be changed to any other

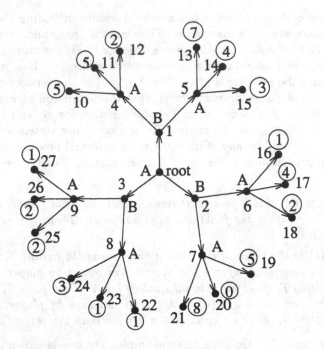

FIGURE 7.1. An example for the algorithms.

values without affecting the final result of the computation. A similar thing happens in the computation of $\pi(3)$. Here, $\pi(8) = 3$ and so it is assured that $\pi(3) \le 3$. Thus, $\sigma(r) \ne 3$ and therefore the computation of $\pi(9)$ is unnecessary. The naive algorithm has no way of taking advantage of the observations we have just made.

The second algorithm is an improvement in that it can stop itself from doing this sort of unneeded calculations. The idea is that it keeps track of a variable which represents a lower bound for the value of $\pi(r)$. When a vertex belonging to B is under consideration, the computation is stopped when it is seen that no possibility exists of beating the lower bound. When a vertex belonging to A is under consideration, the value of this lower bound can be adjusted upward when it is seen that a larger value can be guaranteed.

ALGORITHM 7.2 (BRANCH AND BOUND). *We are given a game tree T as just described. An optimal move from the root for the player owning it is computed.*

(1) *We define two functions $\pi(u, \alpha)$ and $\sigma(u, \alpha)$; $\pi(u, \alpha)$ is defined on the set of all vertices u and all real numbers α, and is real-valued;*

$\sigma(u,\alpha)$ *is defined on the set of all nonterminal vertices u and all real numbers α, and takes its values in the set of vertices. They are defined recursively as follows.*

(a) *If u is terminal, let*

$$\pi(u,\alpha) = p_A(u) \text{ for all } \alpha.$$

(b) *If u belongs to A, let $\{v_1, \ldots, v_d\}$ be the children of u and compute as follows.*

   (i) *Set $i = 1$, $\gamma = \alpha$, $s = -\infty$, $w = v_1$.*

   (ii) *Let $t = \pi(v_i, \gamma)$.*

   (iii) *If $t > s$ set $s = t$, $w = v_i$.*

   (iv) *If $s > \gamma$, set $\gamma = s$.*

   (v) *Set $i = i + 1$.*

   (vi) *If $i \leq d$ go to (ii).*

   (vii) *Set $\pi(u,\alpha) = s$ and $\sigma(u,\alpha) = w$.*

(c) *If u belongs to B, let $\{v_1, \ldots, v_d\}$ be the children of u and compute as follows.*

   (i) *Set $i = 1$, $s = +\infty$, $w = v_1$.*

   (ii) *Let $t = \pi(v_i, \alpha)$.*

   (iii) *If $t < s$ set $s = t$, $w = v_i$.*

   (iv) *If $s \leq \alpha$, go to (vii).*

   (v) *Set $i = i + 1$.*

   (vi) *If $i \leq d$ go to (ii).*

   (vii) *Set $\pi(u,\alpha) = s$ and $\sigma(u,\alpha) = w$.*

(2) *The optimal move from the root for the player owning it is*

$$\sigma(r, -\infty).$$

The key difference between this algorithm and the naive algorithm is found at the point where the variable s is compared to α. This occurs when the vertex under consideration belongs to B and results in a "cutoff" of the computation if $s \leq \alpha$. The new algorithm has to do a little more work at each vertex (in order to update the variable γ, and to decide whether to cut off computation) but has to consider fewer vertices.

7.1.3. *The Alpha-Beta Pruning Algorithm*

An additional saving in computation can be made. In the example of Figure 7.1, consider the computation of $\pi(1)$. After $\pi(4) = 5$ has been computed, we see that $\pi(1) \leq 5$. In the computation of $\pi(5)$, $\pi(13) = 7$ is computed first and tells us that $\pi(5) \geq 7$. Thus, $\pi(1)$ is known to be 5. Therefore, the rest of the calculation of $\pi(5)$ can be cut off. The branch-and-bound algorithm is not able to take advantage of such a "deep cutoff," but the third algorithm is. The idea is to keep track of a second variable which plays a role symmetric to that played by the variable used in the branch-and-bound algorithm. The new variable is an *upper* bound on the value of the game. It may be revised downward when a vertex belonging to player B is under consideration. When a vertex owned by A is under consideration, the computation is stopped if it is seen to be impossible to return a value less than this upper bound.

ALGORITHM 7.3 (ALPHA-BETA PRUNING). *We are given a game tree T as before. An optimal move from the root for the player owning it is computed.*

(1) *We define two functions $\pi(u, \alpha, \beta)$ and $\sigma(u, \alpha, \beta)$; $\pi(u, \alpha, \beta)$ is defined on the set of all vertices u and all real numbers α and β, and is real-valued; $\sigma(u, \alpha, \beta)$ is defined on the set of all nonterminal vertices u and all real numbers α and β, and takes its values in the set of vertices. They are defined recursively as follows.*

  (a) *If u is terminal, let*

$$\pi(u, \alpha, \beta) = p_A(u) \text{ for all } \alpha \text{ and } \beta.$$

  (b) *If u belongs to A, let $\{v_1, \ldots, v_d\}$ be the children of u and compute as follows.*
   (i) *Set $i = 1$, $\gamma = \alpha$, $s = -\infty$, $w = v_1$.*
   (ii) *Let $t = \pi(v_i, \gamma, \beta)$.*
   (iii) *If $t > s$ set $s = t$, $w = v_i$.*
   (iv) *If $s \geq \beta$ go to (viii).*
   (v) *If $s > \gamma$ set $\gamma = s$.*
   (vi) *Set $i = i + 1$.*
   (vii) *If $i \leq d$ go to (ii).*
   (viii) *Set $\pi(u, \alpha, \beta) = s$ and $\sigma(u, \alpha, \beta) = w$.*
  (c) *If u belongs to B, let $\{v_1, \ldots, v_d\}$ be the children of u and compute as follows.*
   (i) *Set $i = 1$, $\delta = \beta$, $s = +\infty$, $w = v_1$.*
   (ii) *Let $t = \pi(v_i, \alpha, \delta)$.*
   (iii) *If $t < s$ set $s = t$, $w = v_i$.*
   (iv) *If $s \leq \alpha$ go to (viii).*
   (v) *If $s < \delta$ set $\delta = s$.*
   (vi) *Set $i = i + 1$.*
   (vii) *If $i \leq d$ go to (ii).*
   (viii) *Set $\pi(u, \alpha, \beta) = s$ and $\sigma(u, \alpha, \beta) = w$.*
(2) *The optimal move from the root for the player owning it is*

$$\sigma(r, -\infty, +\infty).$$

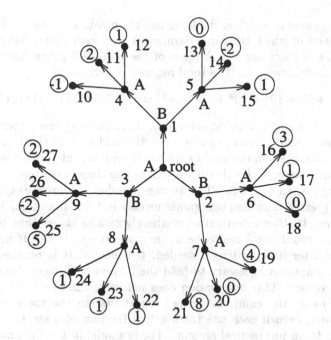

FIGURE 7.2. Game tree for exercises.

These three algorithms are discussed in detail in [KM75]. Proofs are given there that they do what we have claimed they do. Also, some estimates are made of the computational efficiency of alpha-beta pruning in comparison to the naive algorithm. There are also discussions of alpha-beta pruning and its application to specific games in [Sol84] and [Lev84].

Exercises

(1) Apply the naive algorithm to the game in Figure 7.2. Compute the values of $\pi(u)$ and $\sigma(u)$ for every vertex u.

(2) Apply the branch-and-bound algorithm to the game in Figure 7.2. Compute $\pi(r, -\infty)$ and $\sigma(r, -\infty)$. Indicate where cutoffs occur.

(3) Apply alpha-beta pruning to the game in Figure 7.2. Compute $\pi(r, -\infty, +\infty)$ and $\sigma(r, -\infty, +\infty)$. Indicate where deep cutoffs occur.

7.2. Evaluation Functions

For serious games, the full game tree is far too large to be analyzed. In order to get an intuitive feeling for the sizes involved, let us consider an imaginary game in which each player has ten possible moves at each of her vertices, and in which the game terminates after each player has made ten moves. Then there are ten children of the root, 100 grandchildren, 1000 great-grandchildren, etc. The total number of vertices is

$$1 + 10 + 10^2 + 10^3 + \cdots + 10^{20} = 111111111111111111111.$$

Even if the information about one vertex could be compressed into one byte of computer memory, no computer on earth could contain this tree. In fact, if all the computers in the world were wired together into one, it would still not be large enough. Furthermore, games like chess have trees which are much larger than this one. The average number of moves available to a player is greater than ten, and games usually last for more than 20 moves.

It is true that an examination of the algorithms shows that the entire game tree would not have to be in memory at one time. If memory is reclaimed after it is no longer needed, the most that is required at one time is the amount necessary to hold the longest path from the root to a terminal vertex. This observation does not solve the problem. The time required to do the computation is proportional to the total number of vertices and, even if each one takes a tiny fraction of a second, the total builds up to an impractical amount. For example, in the imaginary game just discussed, suppose that the naive algorithm is used and that the time required to compute $\pi(u)$ is only 10^{-9} seconds (for each u). Then the total time required is more than 10^{11} seconds, which is over 3000 years. If we had a computer 10 times as fast, and if the alpha-beta pruning algorithm cut off 90 percent of the vertices, then the calculation could be carried out in only 30 years.

If the algorithms just discussed are to be of any use, they will have to be applied to trees smaller than full game trees.

7.2.1. *Depth-Limited Subgames*

Let T be a game tree and let u be a nonterminal vertex of T. Suppose, to be definite, that u belongs to player A. If one of the three algorithms could be applied to the cutting T_u, then the result would be an optimal move for A from u. But if the entire tree T is enormous, then most of the cuttings T_u are too large to analyze. To obtain a suitably small tree, we use the following definition.

DEFINITION 7.1. Let T be a game tree and let u be a nonterminal vertex of T. Also let m be a positive integer. The *depth-limited subgame* $S(u, m)$ with root u and depth m consists of those vertices v of T_u such that the path from u to v has length at most m. The edges of $S(u, m)$ are the edges of T which begin and end at vertices of $S(u, m)$.

Thus, a depth-limited subgame $S(u, m)$ consists of all vertices of the game tree which can be reached in m or fewer moves, starting at u. In other words, its vertices are the states of the game which player A can "see" by looking ahead at most m moves. The terminal vertices of $S(u, m)$ are, in general, not terminal in T. However, by assigning numbers to these vertices, we can make $S(u, m)$ into a game tree. If m is not too large, $S(u, m)$ will be of reasonable size and we can apply the algorithms to it. If, further, the assigned numbers reasonably reflect the values of their vertices (to A), then the move produced by the algorithm should be a good one (although perhaps not the best one).

Let us now make a definition. A procedure for assigning values to the vertices of the game tree (that is, to the possible states of the game) is called an *evaluation function*. Let $e(u)$ denote such a function. We assume that the evaluation is from A's point of view, and that the evaluation of u from B's point of view is $-e(u)$ [this ensures that $S(u, m)$ is zero-sum]. Second, if u happens to be terminal in T, then it is reasonable to require that

$$e(u) = p_A(u).$$

The idea of an evaluation function is that it assigns a real number to every state of the game. This number is a measure of the worth to player A of that state of the game. For example, if u is a state of the game from which A should (by playing correctly) be able to gain a large payoff, then $e(u)$ will be relatively big.

Observe that, in theory, there is a way to define a perfect evaluation function— simply let $e(u) = \pi(u)$ be the value of T_u, as computed by, for example, the alpha-beta pruning algorithm! As we have been saying, this evaluation function is usually not computable and so we seek one which, though imperfect, is computable and is an approximation to $\pi(u)$. The actual design of such a realistic evaluation function is a tricky business. It clearly requires a good deal of knowledge of the game being studied. For an interesting discussion of the evaluation of chess positions, [Lev84] is recommended. A game-playing program is likely to spend a great part of its time computing values of the evaluation function. Therefore, it is a good idea to make that function as simple as possible. The goals of simplicity and nearness of approximation to $\pi(u)$ are antagonistic, and part of the problem in writing game-playing programs is to strike a balance between them.

In summary, the general procedure which a game-playing program goes through is this. It is given a depth m (either fixed, or set by the user); when it is the program's turn to move (from vertex u, say), it applies the alpha-beta pruning algorithm to the depth-limited subgame $S(u, m)$ (which has been made into a game tree via an evaluation function). The result is a move which may not be optimal, but which, one hopes, is good. It is clear that the quality of the program's play depends on the quality of the evaluation function and that it improves as m increases.

The task of writing a program to play a specific games involves solving many problems which depend on the game. The way in which the current game situation is represented internally, how moves are encoded, and so forth, are all highly game-dependent. Moreover, knowledge of the game often allows the programmer to introduce extra efficiency into the program—a move which a skilled player knows will probably lead to trouble can be avoided. In general, writing a game-playing program which plays as well as a mediocre player is not so hard, but writing one which plays at a championship level is a project which involves as much art as science.

7.2.2. *Mancala*

We will discuss in some detail a game which is complicated enough to be nontrivial to play, but which is such that an evaluation function is easily written. It is one of a very large and very old family of board games which bears the generic name *mancala*. These games are played in all parts of Africa and in the Middle East and Asia. Their origins are lost in history. See [Mur52] and [Bel60] for more information about them.

The game we discuss here may not be exactly like any of the versions of mancala actually played. It is very close to the Indonesian game congklak discussed in [BC88]. The board for our game consists of two rows of six holes, together with two larger bowls (called *stores*). There are also 72 small hard objects—seeds, marbles, or beans would all be suitable, but we refer to them as beads. At the start of play, each of the 12 holes contains 6 beads, and both stores are empty. The mancala board in its initial state is shown in Figure 7.3. The holes numbered 1 through 6 belong to player A, those numbered 7 through 12 belong to B, and each player owns a store (as indicated). Player A moves first. Her move consists of choosing one of the holes on her side of the board, removing the beads contained in it, and "sowing" them in a clockwise direction (as indicated by the arrows). To sow the beads means to add one to each hole in turn (starting with the hole following the chosen one, and including the player's own store but not her opponent's). If the last bead is sown in the player's store, then the move is over and the other player moves. If the last bead is sown in a hole which is "loaded" (that is, nonempty), then its contents are sown (and so forth, as long as the last bead is sown in a loaded hole). If the last bead is sown in an empty hole, then the move is over. However, if this empty hole belongs to the player whose move it is, then whatever beads are in the opposite hole (which belongs to the opponent) are moved to the player's store. The players alternate in this way until a player is unable to move

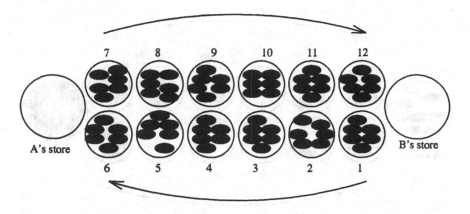

FIGURE 7.3. Initial configuration of a mancala board.

(because all his holes are empty). At this point, the other player puts all the beads remaining in her holes into her store. The winner is the player with more beads at the end.

For example, suppose A's first move is to sow from hole number 1. Then one bead goes into each of holes 2 through 6, and the last bead goes into A's store. A's turn is then over, and B moves. For another example, suppose A's first sowing is from hole 2. Then holes 3 through 6 receive an additional bead, A's store receives one, and hole 7 receives the last one. Since this last hole is loaded, its contents are then sown. The last one goes into the empty hole 2. The move is over, except that the 7 beads in the opposite hole 11 go into A's store. Thus, at the end of this move, A's store contains 8 beads. Of course, B's store contains none. Figure 7.4 shows the state of the board after A has made the move just described, and B has sown from hole 12.

There is an obvious evaluation function for mancala. Let u denote a vertex of the game tree. Then u corresponds to a certain state of the game, that is, knowing that we are at u tells us how many beads there are in each hole, how many there are in each player's store, and whose turn it is to move. We define $e(u)$ to be the number of beads in A's store minus the number in B's. For example, if u is the root, then $e(u) = 0$; if u is the vertex reached after A sows from hole 2, then $e(u) = 8$; if u is the vertex after A sows from hole 2 and B sows from hole 12, then $e(u) = -13$.

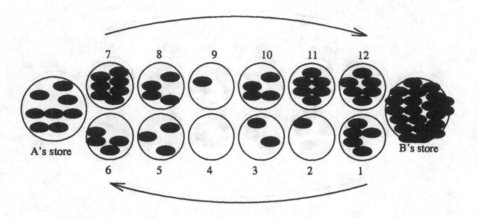

FIGURE 7.4. Mancala board after two moves.

A computer program was written to play mancala. It uses the evaluation function just described, and the depth can be set by the user. We now discuss what was learned about mancala and about the three algorithms from experimenting with this program. First of all, the program plays very well. Even with the depth set at 2, human players (who were, however, not experts at the game) found it almost impossible to win. It is an interesting question whether either player has an advantage in mancala [see, also, Exercise (3)]. The experiment was tried of having the program play against itself for various choices of the depth. The result was always that player A wins. For example, the score was 42 to 30 when the depth was set at 6. These results make it extremely likely that A has the advantage.

The comparisons among the three algorithms illustrate the striking superiority of branch-and-bound over naive, and of alpha-beta pruning over branch-and-bound. As just discussed, the program played against itself for various choices of the depth. The program was written in three different ways, implementing the three algorithms. Table 7.1 shows the total number of calls to the functions $\pi(u)$, $\pi(u, \alpha)$, and $\pi(u, \alpha, \beta)$, respectively, for three different values of the depth.

TABLE 7.1. Comparison of the algorithms.

Depth	Naive	Branch and Bound	Alpha-Beta
2	577	489	311
4	10944	5323	3695
6	212165	49523	25245

7.2.3. Nine-Men's Morris

Our second example is a member of another ancient family of games which bears the generic name *morris*[1]. Besides the version discussed here (nine-men's morris), there are also three-men's, six-men's, and twelve-men's variants. The board for the game is shown in Figure 7.5. To illustrate the great antiquity of these games, we mention that this same board has been found incised on a roof slab of the temple of Kurna built at Thebes on the Nile in ancient Egypt. Construction of this temple was begun in the reign of Rameses I who ruled Egypt from 1400 to 1366 B.C. One pictures the workers on this great project playing the game during their lunch breaks. The games were played in many other parts of the world, and nine men's was popular in Europe through the Middle Ages. It is a shame that it is now almost extinct— the rules and board are simple, but the game is not at all trivial to play. See [Mur52] and [Bel60] for more details.

As seen in Figure 7.5, the board consists of three concentric squares, together with line segments connecting the centers of the squares. The 24 *points* of the board are labeled $a1, \ldots, a8$, $b1, \ldots, b8$, and $c1, \ldots, c8$. Each of the two players has nine small objects. The ones belonging to a given player can be identical, but those belonging to player A must be distinguishable from those belonging to player B. We refer to these objects as *pieces*. The game has two phases. In the first phase, the players take turns placing pieces at points on the board. If a player achieves a *mill*, that is, places three pieces in a row along one of the line segments (for example, $b7 - b8 - b1$ or $a4 - b4 - c4$), then she removes any one of the other player's pieces which is not in a mill. For example, suppose that player A has pieces at points $a7$, $a8$, $a1$, and $a2$; that player B has pieces at points $b4$ and $a4$; and that it is B's turn to play. B can place a piece at $c4$ and then take A's piece at $a2$. The pieces at $a7$, $a8$, and $a1$ cannot be taken because they are in a mill. After both players have placed their nine pieces, the first phase of the game is over. In the second phase, a player's turn consists of moving a piece from a point to an adjacent empty point. For example, if A has a piece at point $c2$ and point $c1$ is empty, she may move her piece there. If such a move causes a mill to form, then any piece belonging to the

---

[1] This name comes from the Low Latin *merellus*, a coin or token.

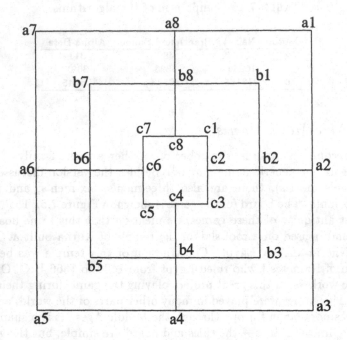

FIGURE 7.5. Board for nine men's morris.

other player which is not in a mill can be removed. The game is over when the player whose turn it is has only two pieces left, or is unable to move because all his pieces are blocked. At this point, the other player wins.

For example, suppose the game has reached the situation shown in Figure 7.6. In this diagram, the round pieces belong to player A and the square ones to player B. It is A's turn to move. She has four possible moves, $c8 - c1$, $c6 - c5$, $b2 - c2$, $a3 - a4$. Here, the notation $c8 - c1$, for example, means to move the piece at $c8$ to the empty point $c1$. Two of these moves seem poor because they would allow player B to form a mill on his next move. In this situation, player A's choice is $b2 - c2$. At a later move, she can choose either $a2 - b2$ or $c2 - b2$, thus forming a mill and gaining a piece.

A computer program was written to play nine-men's morris. The first evaluation function tried was quite simple—it counted the number of A's pieces on the board, and subtracted the number of B's pieces. With this function, the program played at a mediocre level. In fact, it was fairly easy to beat when it played first, even with the depth set at 6. (Setting the depth to a larger value made the program unacceptably slow.) This evaluation function was replaced by one which also counted pieces, but

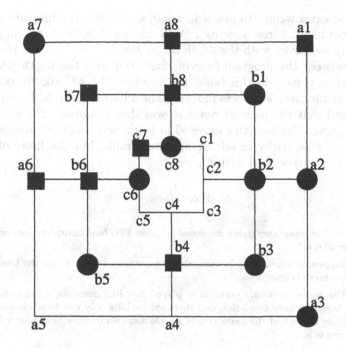

FIGURE 7.6. Nine men's morris board after 18 moves.

TABLE 7.2. A nine-men's morris game.

(1)	$a8$	$a5$
(2)	$a7$	$a1$
(3)	$b8$	$c8$
(4)	$a3$	$c6$
(5)	$b1$	$b7$
(6)	$b2$	$b3$
(7)	$c2$	$a2$
(8)	$c3$	$c1$
(9)	$b4$	$b6$
(10)	$a7 - a6$	$c6 - c7 : a6$
(11)	$a3 - a4$	$a2 - a3$
(12)	$c3 - c4 : b3$	$c7 - c6$
(13)	$b4 - b3 : c6$	$a5 - a6$
(14)	$b3 - b4 : a6$	$b6 - c6$
(15)	$b4 - b3 : c6$	$b7 - b6$
(16)	$b3 - b4 : a3$	$b6 - c6$
(17)	$b4 - b3 : c6$	$c8 - c7$
(18)	$b3 - b4 : c1$	

which gave extra weight to pieces in a mill and to pieces which are mobile (that is, not blocked from moving). With this function, the program played reasonably well, even with the depth set to 3 or 4. Table 7.2 is a record of a game between the program (moving first, and with the depth set at 4) and a human player. In this table, the notation $b4 - b3 : c6$, for example, means that the piece at $b4$ was moved to the adjacent point $b3$ (completing a mill), and that the piece at point $c6$ was then removed. The computer won the game. The human's move $b6$ in line 9 was a crucial mistake. He was trying to be tricky in setting up future mills, but the more obvious move $c7$ would have been better.

Exercises

(1) For the imaginary game discussed on page 191, how many vertices are there in $S(r, 4)$?

(2) Suppose A's first move in mancala is to sow from hole 3. Draw the board after her move is complete.

(3) The game of *simple mancala* is played just like mancala, except that each player has only two holes, and there are initially only two beads in each hole. Draw enough of the game tree of simple mancala to show that player B has a sure win.

Appendix
Solutions

In this appendix, we present solutions to some of the exercises. The notation (n.m.p) stands for exercise number p at the end of Section m of Chapter n.

(1.1.1) There are 10 vertices and, therefore, 10 cuttings. One of them is the tree itself. Two others are shown in Figure A.1.

(1.1.2) There are 10 quotient trees. Six of them (the ones corresponding to the terminal vertices) are the tree itself. One of the others is shown in Figure A.2.

(1.1.3) There are six terminal vertices. Each nonempty set of terminal vertices uniquely determines a subtree (by Theorem 1.4). Thus, the number of subtrees is

$$2^6 - 1 = 63.$$

(1.1.9) Each edge (u, v) contributes $+1$ and -1 to the sum [$+1$ to $\rho(u)$ and -1 to $\rho(v)$]. Thus

$$\sum_{u \in V(G)} \rho(u) = \sum (1 - 1) = 0,$$

where the second sum is over all edges.

(1.1.10) From Exercise 1.1.9, we know that

$$\sum_{u \in V(D)} \rho(u) = 0.$$

Let

$$V_1 = \{u : \rho(u) \text{ is odd}\},$$

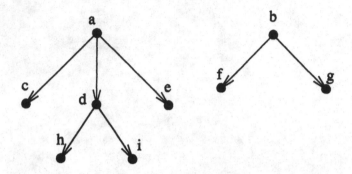

FIGURE A.1. Two cuttings.

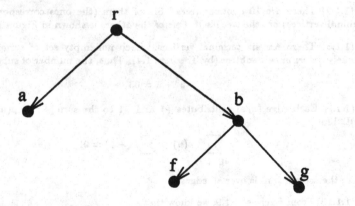

FIGURE A.2. A quotient tree.

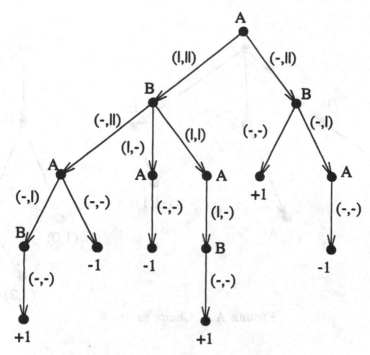

FIGURE A.3. A tree for very simple nim.

and
$$V_2 = \{u : \rho(u) \text{ is even}\}.$$

Then
$$\sum_{V_1} \rho(u) + \sum_{V_2} \rho(u) = \sum_{V_1 \cup V_2} \rho(u) = 0.$$

The sum over V_2 is even, and, therefore, so is the sum over V_1. Since each term of this sum is odd, we must have that the number of terms is even.

(1.2.4) B has only two strategies, L and R. They lead to the same payoff if A moves L or R. However, if A moves M, a payoff of 2 is possible from B's moving R, and not from moving L. Thus, B prefers R. A's best initial move is M (and either second move).

(1.2.6) Figure A.3 shows a game tree for this version of nim. The designations on the edges show the state of the piles of matches after the move. For example, the designation $(|, ||)$ means that, on completion of that move, there will be one pile with one match, and one with two. An empty pile is denoted "–." The terminal vertices are denoted either +1 (a win for player A) or −1 (a win for player B).

Player B has a sure win. If A moves left from the root, B moves middle; if A moves right, B moves right.

(1.3.1) Player A has three strategies, R, LL, LR. Strategy R means that he moves right from the root (and then has no further choices). LL and LR mean that he moves

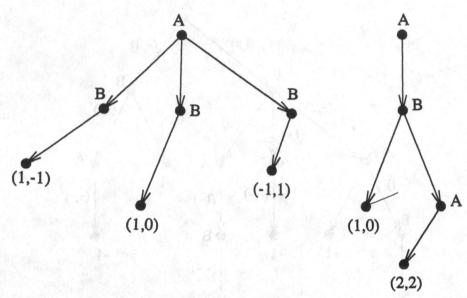

FIGURE A.4. Choice subtrees.

left from the root and moves left and right, respectively, if he reaches the other vertex where a choice exists. Player B has six strategies, LL, LR, ML, MR, RL, and RR. For example, MR means that if A moves left from the root, B moves middle, while if A moves right from the root, B moves right.

(1.3.6) Player A has four choice subtrees, L, R, ML, and MR. Here, MR, for example, means that A moves middle from the root, and moves right if B moves right. All four of these are strategies for A since A has perfect information. Player B has eight choice subtrees since she has two possible moves at each of her three vertices. Only two of these are strategies because she must make the same move at each of the three vertices. We denote these by L and R. One choice subtree for each player (L for B, and ML for A) is shown in Figure A.4.

(1.4.2) Player A has four strategies, LL, LR, RL, RR. Here, for example, LR means that she moves left from either of the two vertices owned by her which are children of the root, and that she moves right from either of the two vertices in her other information set. Player B has eight strategies, xyz, where x, y, and z can be L or R. For example, RLL means that he moves right from the child of the root belonging to him, moves left from either of the two vertices in the left-hand one of the other two information sets, and moves left from either of the two vertices in his remaining information set.

(1.5.1) Trying all $2 \times 4 = 8$ combinations, we find that there are four equilibrium pairs

$$(L, L), (ML, R), (MR, L), (MR, R).$$

(1.6.2) Player A's payoff matrix is

$$\begin{pmatrix} 1 & -1 \\ -1 & 1 \\ 1 & 2 \\ 1 & 2 \end{pmatrix},$$

while B's payoff matrix is

$$\begin{pmatrix} -1 & -1 \\ 1 & 1 \\ 0 & 2 \\ 0 & 0 \end{pmatrix}.$$

Here, the rows are in the order L, R, ML, MR; the columns are in the order L, R.

(1.6.3) After tedious calculation, we get the payoff matrices. For player A, it is

	LLL	LLR	LRL	RLL	LRR	RLR	RRL	RRR
LL	0	1/4	1/4	0	1/2	1/4	1/4	1/2
LR	-1/2	-1/4	-1/4	0	0	1/4	1/4	1/2
RL	1	3/4	5/4	1	1	3/4	5/4	1
RR	1/2	1/4.	3/4	1	1/2	3/4	5/4	1

For player B, the payoff matrix is

	LLL	LLR	LRL	RLL	LRR	RLR	RRL	RRR
LL	-1/2	0	-1/2	1	0	3/2	1	3/2
LR	0	1/2	1/2	0	1/2	1/2	0	1/2
RL	-1/4	0	-1/4	5/4	0	3/2	5/4	3/2
RR	1/4	1/2	1/4	1/4	1/2	1/2	1/4	1/2

For example, the entries corresponding to strategies (RL, RLL) are computed as follows: If the chance move is to the left, the terminal vertex reached has payoffs $(1,1)$; if the chance move is to the middle, the terminal vertex reached has payoffs $(1,0)$; if the chance move is to the right, the payoffs for the terminal vertex reached are $(1,2)$. Thus, the pair of expected payoffs is

$$(1/4)(1,1) + (1/4)(1,0) + (1/2)(1,2) = (1, 5/4).$$

Hence, 1 appears in the (RL, RLL) entry in the first matrix, and 5/4 in the (RL, RLL) entry in the second matrix.

The equilibrium pairs can be found from the two matrices. They are

$$(RL, RLR), (RL, RRR), (RR, RLR), (RR, RRR).$$

(1.6.8) (a) Each player has two strategies: 1 (that is, hold up one finger), and 2 (that is, hold up two fingers).

(b) The payoff matrices for Mary and Ned, respectively, are

$$\begin{pmatrix} 1 & -2 \\ -1 & 2 \end{pmatrix}, \quad \begin{pmatrix} -1 & 2 \\ 1 & -2 \end{pmatrix}.$$

(c) A pair of strategies is in equilibrium if the corresponding entry in Mary's matrix is a maximum in its column, while the corresponding entry in Ned's matrix is a maximum in its row. From the two matrices, we see that there are no pairs of strategies satisfying this condition.

(1.6.10) The two equilibrium 3-tuples are $(1, 2, 1)$ and $(3, 1, 1)$.

(2.1.1) The entry 0 in row 2, column 2, is the only saddle point.

(2.1.3) We have:

- In order for -2 to be a saddle point,

$$-2 \leq a \text{ and } -2 \geq a.$$

Thus, $a = -2$.
- In order for 1 to be a saddle point,

$$1 \leq a \text{ and } 1 \geq a.$$

Thus, $a = 1$.
- In order for the upper right-hand corner to be a saddle point,

$$a \leq -2 \text{ and } a \geq 1.$$

These cannot both be true, and so the upper right-hand corner is never a saddle point.
- In order for the lower left-hand corner to be a saddle point,

$$a \leq 1 \text{ and } a \geq -2.$$

Thus, $-2 \leq a \leq 1$.

In summary, there is a saddle point if $-2 \leq a \leq 1$. If $-2 < a < 1$, the only saddle point is the lower left-hand corner. If $a = -2$, then both entries in the first column (both equal to -2) are saddle points. If $a = 1$, then both entries in the second row (both equal to 1) are saddle points.

(2.1.5) We have

$$\min_j m_{1j} = 0, \min_j m_{2j} = -1, \min_j m_{3j} = 0, \min_j m_{4j} = -1.$$

Thus,

$$u_r(M) = \max_i \min_j m_{ij} = 0.$$

Also,

$$\max_i m_{i1} = 3, \max_i m_{i2} = 2, \max_i m_{i3} = 3, \max_i m_{i4} = 2.$$

Thus

$$u_c(M) = \min_j \max_i m_{ij} = 2.$$

(2.2.1) We compute

$$E(1, (2/5, 1/3, 4/15)) = (2/5)(1) + (1/3)(2) + (4/15)(3) = 28/15,$$

$$E(2, (2/5, 1/3, 4/15)) = (2/5)(3) + (1/3)(0) + (4/15)(2) = 26/15,$$

$$E(3, (2/5, 1/3, 4/15)) = (2/5)(2) + (1/3)(1) + (4/15)(0) = 17/15.$$

Thus, the best *pure* strategy for the row player against the given strategy for the column player is to play row 1. There is no *mixed* strategy which does better than this because, if $\vec{p}$ is a mixed strategy for the row player, and $\vec{q} = (2/5, 1/3, 4/15)$, then

$$E(\vec{p}, \vec{q}) = p_1 E(1, \vec{q}) + p_2 E(2, \vec{q}) + p_3 E(3, \vec{q}) \leq E(1, \vec{q}).$$

(2.2.5) We compute

$$E(1, \vec{q}) = 19/52, \quad E(2, \vec{q}) = -12/13, \quad E(3, \vec{q}) = 19/52,$$

and
$$E(4,\vec{q}) = 19/52, \quad E(5,\vec{q}) = 19/52.$$

Also,
$$E(\vec{p},1) = 19/52, \quad E(\vec{p},2) = 19/52, \quad E(\vec{p},3) = 29/26,$$

and
$$E(\vec{p},4) = 19/52, \quad E(\vec{p},5) = 19/52.$$

Thus,
$$\max_i E(i,\vec{q}) = 19/52 \text{ and } \min_j E(\vec{p},j) = 19/52.$$

In order that both $\vec{p}$ and $\vec{q}$ be optimal, we must have $p_2 = 0$ and $q_3 = 0$. They both are, and so both are optimal. Finally, 19/52 is the value of the game because
$$\max_i E(i,\vec{q}) = \min_j E(\vec{p},j) = 19/52.$$

(2.3.1) There are no saddle points. A theorem in the text applies. We compute
$$\pi_1(p) = -p, \quad \pi_2(p) = 2p - 2(1-p) = 4p - 2.$$

Setting these equal, we get $p = 2/5$. Then the value of the game is
$$\pi_1(2/5) = \pi_2(2/5) = -2/5.$$

To get the column player's value, compute
$$\pi^1(q) = -q + 2(1-q) = -3q + 2, \pi^2(q) = -2(1-q) = 2q - 2.$$

Setting these equal gives $q = 4/5$. As a check, compute
$$\pi^1(4/5) = -2/5 = \pi^2(4/5).$$

In summary, the solution is
$$v = -2/5, \quad \vec{p} = (2/5,3/5), \quad \vec{q} = (4/5,1/5).$$

Then, if the column player is playing $(1/3,2/3)$, we compute
$$E(1,(1/3,2/3)) = 1, \qquad E(2,(1/3,2/3)) = -4/3.$$

If the row player plays the optimal strategy $(2/5,3/5)$ against $(1/3,2/3)$, his payoff is
$$(2/5)(1) + (3/5)(-4/3) = -2/5.$$

Thus, he should play row 1 instead.

(2.3.3) We compute $\pi_j(p) = E((p,1-p),j)$ for $j = 1,2,3,4$ and get
$$\pi_1(p) = -2p + 1, \quad \pi_2(p) = 2p - 1, \quad \pi_3(p) = -4p + 2, \quad \pi_4(p) = p - 1.$$

We graph these four functions in Figure A.5, indicate their minimum with a heavier line, and circle the maximum of this minimum. It occurs where $\pi_3(p)$ crosses $\pi_4(p)$. Setting these equal, we get
$$p^* = 3/5, \quad v_r = -2/5.$$

We also see from Figure A.5 that columns 1 and 2 are inactive. This leaves us with a 2×2 problem
$$\begin{pmatrix} -2 & 0 \\ 2 & -1 \end{pmatrix}.$$

This is easily solved to give
$$q^* = 1/5, \qquad v_c = -2/5.$$

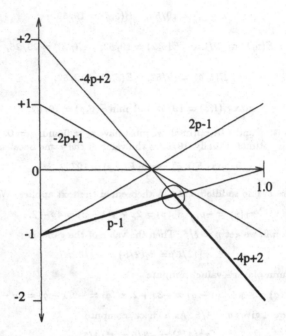

FIGURE A.5. Row value.

In summary, the solution is

$$\vec{p} = (3/5, 2/5), \quad \vec{q} = (0, 0, 1/5, 4/5), \quad v = -2/5.$$

(2.3.4) We compute $\pi^i(q) = E(i, (q, 1 - q))$ for $i = 1, 2, 3, 4$. We get

$$\pi^1(q) = (-1)q + (3)(1 - q) = -4q + 3, \quad \pi^2(q) = 5q - 1,$$

$$\pi^3(q) = -8q + 5, \quad \pi^4(q) = 2q + 1.$$

These are graphed in Figure A.6; their maximum is shown in a heavier line; and the minimum of the maximum is circled. This minimum occurs where $\pi^3(q)$ crosses $\pi^4(q)$. Setting these equal and solving for q gives

$$q^* = 2/5, \quad v_c = 9/5.$$

We see from the figure that rows 1 and 2 are inactive. This leaves us with a 2×2 problem

$$\begin{pmatrix} -3 & 5 \\ 3 & 1 \end{pmatrix}.$$

This is easily solved to give $p^* = 1/5$ and $v_r = 9/5$. In summary, the solution is

$$\vec{p} = (0, 0, 1/5, 4/5), \quad \vec{q} = (2/5, 3/5), \quad v = 9/5.$$

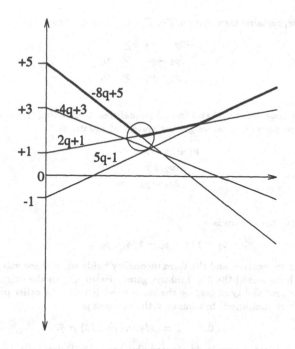

FIGURE A.6. Column value.

(2.3.8) We first deal with the case where there is a saddle point. If

$$1 \leq a \leq 2,$$

then a is a saddle point. Since $a > 0$, the value of the game is positive. If $a \geq 2$, then 2 is a saddle point. Since $2 > 0$, the value is positive. If $a < 1$, there is no saddle point, and we compute optimal mixed strategies. To compute $(p^*, 1 - p^*)$, set

$$ap + 1 - p = 2p - (1 - p),$$

and solve to get

$$p^* = \frac{2}{4-a}, \qquad v_r = 2p^* - (1 - p^*) = \frac{a+2}{4-a}.$$

The row value is zero if and only if $a = -2$. It is negative when $a < -2$, and positive when $a > -2$. Computing the column player's strategy gives

$$q^* = \frac{3}{4-a}, \qquad v_c = \frac{a+2}{4-a} = v_r.$$

(2.4.1) The matrix is skew-symmetric. We notice that row 2 is dominated by row 4. Also, column 2 is dominated by column 4. Erasing the second row and second column, we have

$$\begin{pmatrix} 0 & 2 & -1 \\ -2 & 0 & 1 \\ 1 & -1 & 0 \end{pmatrix}.$$

This, of course, remains skew-symmetric. The three inequalities are

$$-2p_2 + p_3 \geq 0,$$
$$2p_1 - p_3 \geq 0,$$
$$-p_1 + p_2 \geq 0.$$

We arbitrarily choose the first two to make equalities. Combining these with the condition that the p_i's sum to one, we get the system of equations

$$p_1 + p_2 + p_3 = 1,$$
$$-2p_2 + p_3 = 0,$$
$$2p_1 - p_3 = 0.$$

The solution to this system is

$$p_1 = 1/4, \quad p_2 = 1/4, \quad p_3 = 1/2.$$

These are all nonnegative, and the third inequality holds when these values are plugged in. Thus, we have solved the 3×3 matrix game. Returning to the original matrix, we know that the probability of playing the second row is zero; the other probabilities are the numbers just computed. In summary, the solution is

$$v = 0, \qquad \vec{p} = (1/4, 0, 1/4, 1/2) = \vec{q}.$$

(2.4.4) The bottom row is dominated (by row 4). Deleting it leaves the skew-symmetric matrix

$$\begin{pmatrix} 0 & 1 & 2 & -3 \\ -1 & 0 & 1 & 0 \\ -2 & -1 & 0 & 1 \\ 3 & 0 & -1 & 0 \end{pmatrix}.$$

Since the game is symmetric, we try the method of Section 2.4. The inequalities are

$$-r_2 - 2r_3 + 3r_4 \geq 0,$$
$$r_1 - r_3 \geq 0,$$
$$2r_1 + r_2 - r_4 \geq 0,$$
$$-3r_1 + r_3 \geq 0.$$

We first try setting the first three of these to equalities, and attempt to solve the system from those together with the equation

$$r_1 + r_2 + r_3 + r_4 = 1.$$

This system has no solution. Setting the first two inequalities together with the last to be equalities does work. We get

$$(0, 3/4, 0, 1/4).$$

The third inequality (the one not set to an equality) is valid for these values of the unknowns. They are all nonnegative, and so we have a solution. As for the original matrix game, the probability of playing the bottom row must be zero. The solution is

$$\vec{p} = (0, 3/4, 0, 1/4, 0), \quad \vec{q} = (0, 3/4, 0, 1/4), \quad v = 0.$$

If we set the first and the last two inequalities to equalities, we get

$$\vec{r} = (1/6, 0, 1/2, 1/3).$$

However, the omitted inequality is not valid for these values of the unknowns. Finally, omitting the first inequality gives a second valid solution

$$\vec{p} = (0, 1/2, 0, 1/2, 0), \quad \vec{q} = (0, 1/2, 0, 1/2), \quad v = 0.$$

(3.1.2) The given problem is primal. It is clearly feasible since $(0, 0, 0, 0)$ satisfies the constraints. The dual problem is

$$\text{minimize} \quad y_2$$

$$\text{subject to} \quad y_1 \geq 1$$

$$y_2 \geq -1$$

$$-y_1 \geq -1$$

$$-y_2 \geq 1.$$

The last constraint makes this dual problem infeasible (since y_2 cannot be negative). The infeasibility of the dual does *not* tell us that the primal is unbounded. The implication in Corollary 3.2 only goes in the opposite direction. Returning to the primal, however, we see that if we let $x_4 \to +\infty$, while keeping the other variables equal to zero, we get that the objective function goes to infinity. Thus, the primal is unbounded.

(3.2.1) Introducing two slack variables, x_3 and x_4, we get the basic form

$$\text{maximize} \quad 2x_1 + 3x_2$$

$$\text{subject to} \quad x_1 + x_2 - 10 = -x_3$$

$$x_1 - x_2 - 2 = -x_4.$$

Since there are two unknowns in the primal form and two constraints, the total possible number of basic forms is

$$\binom{2+2}{2} = 6.$$

The basis for the basic form already given is $\{x_3, x_4\}$. To find the basic form whose basis is $\{x_1, x_4\}$, solve the first constraint above for x_1 to get

$$x_1 = -x_2 - x_3 + 10. \tag{A.1}$$

Substituting this into the equation for $-x_4$ gives us

$$-2x_2 - x_3 + 8 = -x_4.$$

Then, substituting (A.1) into the objective function gives

$$x_2 - 2x_3 + 20.$$

Finally, applying a little algebra to (A.1) gives

$$x_2 + x_3 - 10 = -x_1.$$

The new basic form, obtained from the first by interchanging x_1 and x_3, is then

$$\text{maximize} \quad x_2 - 2x_3 + 20$$

$$\text{subject to} \quad x_2 + x_3 - 10 = -x_1$$

$$-2x_2 - x_3 + 8 = -x_4.$$

If we now interchange x_4 and x_2 in this second basic form, we get the basic form

$$\text{maximize} \quad 1/2x_4 - 5/2x_3 + 24$$

$$\text{subject to} \quad 1/2x_4 + 1/2x_3 - 6 = -x_1$$

$$-1/2x_4 + 1/2x_3 - 4 = -x_2.$$

Interchanging x_1 and x_3 in this gives

$$\text{maximize} \quad 3x_4 + 5x_1 - 6$$

$$\text{subject to} \quad x_4 + 2x_1 - 12 = -x_3$$

$$-x_4 - x_1 + 2 = -x_2.$$

Interchanging x_2 and x_1 now gives

$$\text{maximize} \quad -2x_4 + 5x_2 + 4$$

$$\text{subject to} \quad -x_4 + 2x_2 - 8 = -x_3$$

$$x_4 - x_2 - 2 = -x_1.$$

An examination of the five basic forms computed so far shows that the only one missing is the one whose basis is $\{x_2, x_4\}$. This can be obtained from any one of the forms already computed, except the last. It is

$$\text{maximize} \quad -3x_3 - x_1 + 30$$

$$\text{subject to} \quad x_3 + 2x_1 - 12 = -x_4$$

$$x_3 + x_1 - 10 = -x_2.$$

(3.3.2) The initial tableau is

x_1	x_2	x_3	-1		
1	1	1	3	$=$	$-x_4$
-1	1	0	0	$=$	$-x_5$
1	2	3	0	$=$	f

(A.2)

This tableau is feasible but nonoptimal. We choose to pivot in the first column. The top entry is the only positive one and so we pivot so as to interchange x_1 and x_4. The second tableau is

x_4	x_2	x_3	-1		
1	1	1	3	$=$	$-x_1$
1	2	1	3	$=$	$-x_5$
-1	1	2	-3	$=$	f

(A.3)

This is still not optimal. We choose the second column to pivot in. The minimum ratio is in the second row and so we pivot so as to interchange x_2 and x_5. The new tableau is

x_4	x_5	x_3	-1		
1/2	$-1/2$	1/2	3/2	$=$	$-x_1$
1/2	1/2	1/2	3/2	$=$	$-x_2$
$-3/2$	$-1/2$	3/2	$-9/2$	$=$	f

$$(A.4)$$

This is still not optimal. There is only one possible pivot column, the third. There is a tie for choice of pivot row. We arbitrarily choose the first. Thus the next pivot will interchange x_3 and x_1. The result is

x_4	x_5	x_1	-1		
1	-1	2	3	$=$	$-x_3$
0	1	-1	0	$=$	$-x_2$
-3	1	-3	-9	$=$	f

$$(A.5)$$

Still not optimal! There is only one possible pivot—the one which interchanges x_5 and x_2. We get

x_4	x_2	x_1	-1		
1	1	1	3	$=$	$-x_3$
0	1	-1	0	$=$	$-x_5$
-3	-1	-2	-9	$=$	f

$$(A.6)$$

This is, at last, optimal. The solution is thus

$$x_1 = x_2 = 0, \quad x_3 = 3; \max f = 9.$$

It took four pivots to do it. We note that a luckier choice of pivot at the beginning would have allowed us to get the answer in one pivot.

(3.4.2) The initial basic form for this problem is

$$\text{maximize} \quad x_1 + x_2 - x_3 - x_4 \qquad (A.7)$$

$$\text{subject to} \quad x_1 + x_2 + x_3 + x_4 - 8 = -x_5$$

$$x_1 - x_2 - x_3 - x_4 + 1 = -x_6$$

$$x_3 - 4 = -x_7.$$

This basic form is infeasible because of the positive constant term in the constraint equation for $-x_6$. We thus apply the feasibility algorithm. The initial basic form for the auxiliary problem is

$$\text{maximize} \quad -u \qquad (A.8)$$

$$\text{subject to} \quad x_1 + x_2 + x_3 + x_4 - u - 8 = -x_5$$

$$x_1 - x_2 - x_3 - x_4 - u + 1 = -x_6$$

$$x_3 - u - 4 = -x_7.$$

The tableau for this basic form is

x_1	x_2	x_3	x_4	u	-1		
1	1	1	1	-1	8	$=$	$-x_5$
1	-1	-1	-1	-1	-1	$=$	$-x_6$
0	0	1	0	-1	4	$=$	$-x_7$
0	0	0	0	-1	0	$=$	h

(A.9)

We have used h to refer to the auxiliary objective function. According to the feasibility algorithm, we should pivot on the entry in row 2, column 5 (thus interchanging u and x_6). The new tableau is

x_1	x_2	x_3	x_4	x_6	-1		
0	2	2	2	-1	9	$=$	$-x_5$
-1	1	1	1	-1	1	$=$	$-u$
-1	1	2	1	-1	5	$=$	$-x_7$
-1	1	1	1	-1	1	$=$	h

(A.10)

As promised in the algorithm, this tableau is feasible. It is not optimal, however. We choose the second column as the pivot column. There is only one choice of pivot row, and so we pivot so as to interchange x_2 and u. The resulting tableau is

x_1	u	x_3	x_4	x_6	-1		
2	-2	0	0	1	7	$=$	$-x_5$
-1	1	1	1	-1	1	$=$	$-x_2$
0	-1	1	0	0	4	$=$	$-x_7$
0	-1	0	0	0	0	$=$	h

(A.11)

This is optimal and the maximum value of h is zero. Thus the original problem is feasible. Following the algorithm, we now set up a (feasible) tableau for the original problem. It is

x_1	x_3	x_4	x_6	-1		
2	0	0	1	7	$=$	$-x_5$
-1	1	1	-1	1	$=$	$-x_2$
0	1	0	0	4	$=$	$-x_7$
2	-2	-2	1	-1	$=$	f

(A.12)

The bottom row of this tableau is obtained as follows. The objective function in the original problem is

$$f = x_1 + x_2 - x_3 - x_4.$$

The nonbasic variables are now x_1, x_3, x_4, x_6, and so we must write f in terms of these. Thus, we want to get rid of x_2 and replace it with x_6. The second row of the tableau gives us $-x_2$ in terms of the nonbasics. Substituting for x_2 from this equation gives us the desired formula for f. The tableau above is feasible but not optimal. One pivot (interchanging x_6 and x_5) gives us a feasible tableau

x_1	x_3	x_4	x_5	-1		
2	0	0	1	7	$=$	$-x_6$
1	1	1	1	8	$=$	$-x_2$
0	1	0	0	4	$=$	$-x_7$
0	-2	-2	-1	-8	$=$	f

(A.13)

From this, we read off the solution

$$x_1 = x_3 = x_4 = 0, \quad x_2 = 8, \quad \max f = 8.$$

(3.5.3) The primal problem for this dual is

$$\text{maximize} \quad -x_2$$

$$\text{subject to} \quad -x_1 \leq 1$$

$$x_1 - x_2 \leq -1$$

$$x_2 \leq 1.$$

The dual/primal tableau is

	x_1	x_2	-1		
y_1	-1	0	1	$=$	$-x_3$
y_2	1	-1	-1	$=$	$-x_4$
y_3	0	1	1	$=$	$-x_5$
-1	0	-1	0	$=$	f
	$= y_4$	y_5	$= g$		

(A.14)

This tableau is feasible as a dual tableau, but infeasible as a primal one. Therefore, it seems easier to use the dual simplex algorithm. It calls for us to pivot so as to interchange y_2 and y_5 (and, thus, x_2 and x_4). The result of this pivot is

	x_1	x_4	-1		
y_1	-1	0	1	$=$	$-x_3$
y_5	-1	-1	1	$=$	$-x_2$
y_3	1	1	0	$=$	$-x_5$
-1	-1	-1	1	$=$	f
	$= y_4$	$= y_2$	$= g$		

(A.15)

This is feasible and optimal, both as a dual and a primal tableau. We read off the solutions

$$y_1 = y_3 = 0, \quad y_2 = 1; \quad \min g = -1.$$

$$x_1 = 0, x_2 = 1, \quad x_3 = 1; \quad \max f = -1.$$

(4.1.1) First, the smallest entry in the matrix is -3. We therefore choose $c = 4$ in the preliminary step. Adding this quantity to each entry gives the modified matrix

$$\begin{pmatrix} 1 & 6 & 4 \\ 5 & 2 & 3 \\ 3 & 4 & 6 \\ 5 & 5 & 1 \end{pmatrix}$$

The initial primal/dual tableau can then be set up. It is

	x_1	x_2	x_3	-1		
y_1	1	6	4	1	$=$	$-x_4$
y_2	5	2	3	1	$=$	$-x_5$
y_3	3	4	6	1	$=$	$-x_6$
y_4	5	5	1	1	$=$	$-x_7$
-1	1	1	1	0	$=$	f
	$= y_5$	$= y_6$	$= y_7$	$= g$		

We choose to pivot in the first column. There is then a choice of two possible pivot rows (2 and 4). We choose row 2. That is, we will interchange x_1 and x_5 (and y_2 and y_5). After the pivot, the new tableau is

	x_5	x_2	x_3	-1		
y_1	$-1/5$	28/5	17/5	4/5	$=$	$-x_4$
y_5	1/5	2/5	3/5	1/5	$=$	$-x_1$
y_3	$-3/5$	14/5	21/5	2/5	$=$	$-x_6$
y_4	-1	3	-2	0	$=$	$-x_7$
-1	$-1/5$	3/5	2/5	$-1/5$	$=$	f
	$=y_2$	$=y_6$	$=y_7$	$=g$		

We choose to pivot in the second column. Then, the only possible pivot row is 4. Thus, we interchange x_2 and x_7 (and y_4 and y_6). The new tableau is

	x_5	x_7	x_3	-1		
y_1	5/3	$-28/15$	107/15	4/5	$=$	$-x_4$
y_5	1/3	$-2/15$	13/15	1/5	$=$	$-x_1$
y_3	1/3	$-14/15$	91/15	2/5	$=$	$-x_6$
y_6	$-1/3$	1/3	$-2/3$	0	$=$	$-x_2$
-1	0	$-1/5$	4/5	$-1/5$	$=$	f
	$=y_2$	$=y_4$	$=y_7$	$=g$		

This is still not optimal. There is only one possible pivot. It is the one which interchanges x_3 and x_6 (and y_3 and y_7). The new tableau is

	x_5	x_7	x_6	-1		
y_1	x	x	x	30/91	$=$	$-x_4$
y_5	x	x	x	1/7	$=$	$-x_1$
y_7	x	x	x	6/91	$=$	$-x_3$
y_6	x	x	x	4/91	$=$	$-x_2$
-1	$-4/91$	$-1/13$	$-12/91$	$-23/91$	$=$	f
	$=y_2$	$=y_4$	$=y_3$	$=g$		

This tableau is optimal and so we have not bothered to compute the irrelevant entries. They are replaced by x. The solution to the primal and dual problems are thus

$$x_1 = 1/7, \quad x_2 = 4/91, \quad x_3 = 6/91$$

$$y_1 = 0, \quad y_2 = 4/91, \quad y_3 = 12/91, \quad y_4 = 1/13$$

$$\max f = \min g = 23/91.$$

According to the change-of-variable formulas, the probabilities for the column player are given by

$$q_1 = (1/7)/(23/91) = 13/23, \quad q_2 = (4/91)/(23/91) = 4/23,$$

$$q_3 = (6/91)/(23/91) = 6/23.$$

Thus, the column player's optimal strategy is $(13/23, 4/23, 6/23)$. Similarly, the row player's optimal strategy is computed to be $(0, 4/23, 12/23, 7/23)$. The value of the *modified* matrix is 91/23. Therefore, the value of the original game is

$$v = 91/23 - c = 91/23 - 4 = -1/23.$$

It is slightly favorable to the column player.

(4.2.3) Each of the two players (call them P_1 and P_2) chooses one of the four vertices A, B, C, D. If they both choose the same vertex, the payoff is zero. If there is an edge from P_1's choice to P_2's choice, then P_1 wins 1 and P_2 loses 1. If there is an edge going the other way, then the payoff also goes the other way. If there is no edge between the choices, then the payoff is zero. The matrix for the game is

$$\begin{pmatrix} 0 & 0 & 0 & 1 \\ 0 & 0 & 1 & -1 \\ 0 & -1 & 0 & 1 \\ -1 & 1 & -1 & 0 \end{pmatrix}.$$

Here, the order of the rows and columns corresponds to the alphabetical ordering of the vertices. This is, of course, a symmetric game and can be solved by the method of Chapter 2. However, an inspection of the matrix shows that the entry in row 1, column 1, is a saddle point. Thus, both players should always choose vertex A.

(4.2.7) For each of the four columns of the matrix for Colonel Blotto's game, compute the expected payoff when that column is played against Colonel Blotto's mixed strategy. We find that these payoffs are 2 for both the first two columns, and 3/2 for the other two. It follows that Attila should put two regiments at one position and the other at the remaining position. The payoff, 3/2, is slightly better for Attila than the value of the game (which was computed to be 14/9). Thus Blotto is punished for not respecting game theory. It should also be noticed that any weighted average of the two pure strategies $(2,1)$ and $(1,2)$ would also give the same payoff.

(4.2.11) There are nine combinations of hands:

- Each is dealt HH(probability = 1/16)—Rose bets, Sue sees, payoff is zero.
- Rose is dealt HH, Sue is dealt LL(probability = 1/16)—Rose bets, Sue folds, payoff is 1.
- Rose is dealt HH, Sue is dealt HL(probability = 1/8)—Rose bets, Sue folds, payoff is 1.
- Rose is dealt LL, Sue is dealt HH(probability = 1/16)—Rose checks, Sue bets, Rose folds, payoff is −1.
- Both are dealt LL(probability = 1/16)—Rose checks, Sue bets, Rose folds, payoff is −1.
- Rose is dealt LL, Sue is dealt HL(probability = 1/8)—Rose checks, Sue bets, Rose folds, payoff is −1.
- Rose is dealt HL, Sue is dealt HH(probability = 1/8)—Rose bets, Sue sees, payoff is −3.
- Rose is dealt HL, Sue is dealt LL(probability = 1/8)—Rose bets, Sue folds, payoff is 1.
- Both are dealt HL(probability = 1/4)—Rose bets, Sue folds, payoff is 1.

Summing the products of these payoffs times the associated probabilities gives, for the expected payoff, the amount −1/16.

(5.1.5) The row player's payoff matrix is

$$\begin{pmatrix} -1 & 1 \\ 2 & 0 \\ 1 & -2 \end{pmatrix}.$$

The row player's maximin value is the value of this matrix game. It is easily computed to be 1/2 (after noting that the third row is dominated by the second).

The column player's payoff matrix is

$$\begin{pmatrix} 3 & 0 \\ -1 & 1 \\ 1 & 1 \end{pmatrix}.$$

To compute the column player's maximin value, we need the *transpose* of this matrix. It is

$$\begin{pmatrix} 3 & -1 & 1 \\ 0 & 1 & 1 \end{pmatrix}.$$

The column player's maximin value is the value of this matrix. The third column is dominated by the second. The value is easily computed to be 3/5.

(5.1.7) We compute

$$\begin{aligned} \pi_1(x,y) &= 2xy - x(1-y) + (1-x)(1-y) \\ &= (4y-2)x - y + 1. \end{aligned}$$

From this, we see that the maximum over x of $\pi_1(x,y)$ is attained for $x = 0$ when $y < 1/2$, for $x = 1$ when $y > 1/2$, and for all x when $y = 1/2$. The set

$$A = \{(x,y) : \pi_1(x,y) \text{ is a maximum over } x \text{ with } y \text{ fixed}\}$$

is shown in Figure A.7 with a solid line.

Then we compute

$$\begin{aligned} \pi_2(x,y) &= -3xy + 3x(1-y) + y(1-x) - 2(1-x)(1-y) \\ &= (-9x+3)y + 5x - 2. \end{aligned}$$

From this, we see that the maximum over y of $\pi_2(x,y)$ is attained for $y = 1$ when $x < 1/3$, for $y = 0$ when $x > 1/3$, and for all y when $x = 1/3$. The set

$$B = \{(x,y) : \pi_2(x,y) \text{ is a maximum over } y \text{ with } x \text{ fixed}\}$$

is shown in Figure A.7. The intersection of the sets A and B is the set of equilibrium pairs. We see from Figure A.7 that there is only one, $(1/3, 1/2)$. We note that the payoffs to the two players corresponding to this equilibrium pair are

$$\pi_1(1/3, 1/2) = 1/2, \pi_2(1/3, 1/2) = -1/3.$$

The maximin values for the row player and column player are the values of the two matrices

$$\begin{pmatrix} 2 & -1 \\ 0 & 1 \end{pmatrix}$$

and

$$\begin{pmatrix} -3 & 1 \\ 3 & -2 \end{pmatrix}.$$

The values are then easily computed to be 1/2 (for the row player) and −1/3 (for the column player).

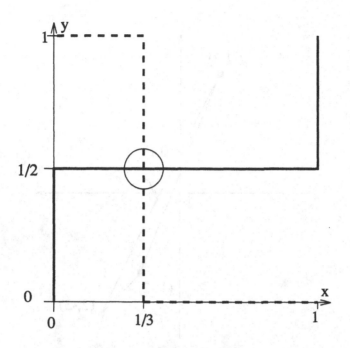

FIGURE A.7. Equilibrium pair.

(5.3.1) The problem concerns the *noncooperative* payoff region. An accurate drawing of this region would allow us to solve the problem, but such drawings are difficult to make. However, the following fact is easily verified from the definition of Pareto optimality: *If a payoff pair (u, v) is in the noncooperative payoff region and is Pareto optimal in the cooperative region, then it is Pareto optimal in the noncooperative region.* The cooperative payoff region is shown in Figure A.8. From it, we immediately see that the three payoff pairs are Pareto optimal.

(5.3.2) From Exercise 5.1.7, the pair of maximin values is $(1/2, -1/3)$. The arbitration pair is to be found in the set of payoff pairs which are Pareto optimal and which dominate $(1/2, -1/3)$. A glance at Figure A.8 shows that this set is the line segment from $(1/2, 0)$ to $(2/3, -1/3)$. It is indicated with a heavier line in Figure A.8 and has an arrow pointing at it. In the terminology of the proof of Theorem 5.4, Case (i) holds. Thus, the arbitration pair occurs at the maximum of the function

$$g(u, v) = (u - 1/2)(v + 1/3),$$

where the payoff pair (u, v) lies in the line segment just mentioned. On this line segment,

$$v = -2u + 1,$$

and

$$1/2 \le u \le 2/3.$$

Thus

$$g(u, v) = (u - 1/2)(-2u + 1 + 1/3) = -2u^2 + 7u/3 - 2/3.$$

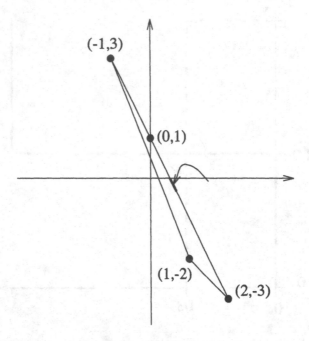

(-1,3)

(0,1)

(1,-2)

(2,-3)

FIGURE A.8. Payoff region (cooperative).

Setting the derivative of this function of u equal to 0 gives

$$u = 7/12 \text{ and so } v = -1/6.$$

This pair is on the line segment and $g(7/12, -1/6) = 1/72$. The values of $g(u, v)$ at both endpoints are zero. We conclude that $(7/12, -1/6)$ is the arbitration pair.

(6.1.2) There are three players. Let us designate them as P_1, P_2, P_3. There are eight coalitions. Besides the grand coalition and the empty coalition, they are

$$\{P_1\}, \{P_2\}, \{P_3\},$$

and

$$\{P_1, P_2\}, \{P_1, P_3\}, \{P_2, P_3\}.$$

The bi-matrix for the two-player game between $\{P_1\}$ and its counter-coalition (that is, $\{P_2, P_3\}$) is

$$\begin{pmatrix} (0,0) & (0,0) & (1,-1) & (0,0) \\ (1,-1) & (2,-2) & (1,-1) & (-1,1) \end{pmatrix}.$$

In this bi-matrix, the rows correspond to the two strategies 1 and 2 for $\{P_1\}$, and the four columns correspond to the joint strategies $(1,1)$, $(1,2)$, $(2,1)$, and $(2,2)$. Thus, the entry in the second row and second column is $(2,-2)$ because the payoff triple corresponding to the strategy combination $(2,1,2)$ is $(2,0,-2)$. From this bi-matrix, the maximin values are quickly computed to give

$$\nu(\{P_1\}) = 0, \qquad \nu(\{P_2, P_3\}) = 0.$$

For the coalition $\{P_2\}$, the bi-matrix is

$$\begin{pmatrix} (0,0) & (-1,1) & (1,-1) & (0,0) \\ (0,0) & (0,0) & (-1,1) & (1,-1) \end{pmatrix}.$$

The rows and columns again correspond to joint strategies in the same order as before. The entry in the first row, third column, is $(1,-1)$ because the payoff triple corresponding to the strategy combination $(2,1,1)$ is $(1,1,-2)$. From the maximin values, we have

$$\nu(\{P_2\}) = -1/3, \qquad \nu(\{P_1, P_3\}) = 1/3.$$

Finally, we get

$$\nu(\{P_3\}) = -2, \qquad \nu(\{P_1, P_2\}) = 2.$$

Also,

$$\nu(\{P_1, P_2, P_3\}) = 0, \qquad \nu(\{\emptyset\}) = 0.$$

We notice that the equation

$$\nu(\mathcal{S}^c) = -\nu(\mathcal{S})$$

holds for every coalition $\mathcal{S}$. The reason is that the game is zero-sum. The characteristic functions of zero-sum games are discussed in a later section.

(6.2.5) The game is zero-sum and essential; thus, by Theorem 6.10, its core is empty. For the fun of it, however, let us verify this by a calculation. Using the values of the characteristic function computed in the previous problem, we see that (x_1, x_2, x_3) is an imputation if

$$x_1 \geq 0, \quad x_2 \geq -1/3, \quad x_3 \geq -2,$$

and

$$x_1 + x_2 + x_3 = 0.$$

Also, an imputation (x_1, x_2, x_3) is in the core if and only if

$$x_1 + x_2 \geq 2, \quad x_1 + x_3 \geq 1/3, \quad x_2 + x_3 \geq 0,$$

since the other inequalities from Theorem 6.7 are automatically true for an imputation. Adding these last three inequalities, we get

$$2(x_1 + x_2 + x_3) \geq 7/3.$$

This contradicts

$$x_1 + x_2 + x_3 = 0,$$

and so the core is empty.

(6.2.6) To verify that ν is a characteristic function, we have to check that superadditivity,

$$\nu(\mathcal{S} \cup \mathcal{T}) \geq \nu(\mathcal{S}) + \nu(\mathcal{T}),$$

holds whenever $\mathcal{S}$ and $\mathcal{T}$ are disjoint coalitions. This is easily checked. For example,

$$\nu(\{P_1, P_2, P_4\}) = 2 \geq -1 + 1 = \nu(\{P_1\}) + \nu(\{P_2, P_4\}).$$

By Corollary 6.8, a 4-tuple (x_1, x_2, x_3, x_4) is in the core if and only if both the following hold:

$$x_1 + x_2 + x_3 + x_4 = \nu(\{P_1, P_2, P_3, P_4\}) = 2,$$

and
$$\sum_{P_i \in S} x_i \geq \nu(S),$$

for every coalition S. It is easily checked that, for example, $(1,0,0,1)$ and $(0,1,0,1)$ satisfy these conditions. Thus the core is not empty.

(6.3.6) Using the formulas in the proof of Theorem 6.14, we compute
$$k = \frac{1}{\nu(\mathcal{P}) - \sum_{i=1}^{N} \nu(\{P_i\})} = \frac{1}{2 - (-2)} = 1/4.$$

Then
$$\begin{aligned}
c_1 &= -k\nu(\{P_1\}) = -(1/4)(-1) = 1/4, \\
c_2 &= 0, \\
c_3 &= 1/4, \\
c_4 &= 0.
\end{aligned}$$

Then
$$\mu(S) = k\nu(S) + \sum_{P_i \in S} c_i$$

defines the $(0,1)$-reduced form of ν. For example,
$$\mu(\{P_1, P_3, P_4\}) = (1/4)\nu(\{P_1, P_3, P_4\}) + (1/4 + 1/4 + 0) = 1/2.$$

(6.4.4) The Shapley value ϕ_i is computed from (6.15) on page 180. For player P_1, the sum is over the coalitions S for which
$$\delta(P_1, S) \neq 0.$$

These are
$$\{P_1\}, \{P_1, P_4\}, \{P_1, P_2, P_3\}, \{P_1, P_2, P_4\}, \{P_1, P_2, P_3, P_4\}.$$

We have
$$\phi_1 = -1/4 + 1/12 + 1/12 + 1/12 + 1/4 = 1/4.$$

The other three are computed in the same way to give the Shapley vector
$$(1/4, 13/12, -1/4, 11/12).$$

It is an imputation (as it is supposed to be). Players P_2 and P_4 are in powerful positions in this game.

Bibliography

[Apo74]   Tom W. Apostol. *Mathematical Analysis*. Addison-Wesley, second edition, 1974.

[Axe84]   Robert M. Axelrod. *The Evolution of Cooperation*. Basic Books, 1984.

[BC88]   Robbie Bell and Michael Cornelius. *Board Games Round the World*. Cambridge University Press, 1988.

[Bel60]   R.C. Bell. *Board and Table Games*. Oxford University Press, 1960.

[Bla77]   R.G. Bland. New finite pivoting rules for the simplex method. *Mathematics of Operations Research*, 2:103–107, 1977.

[BM76]   J.A. Bondy and U.S.R. Murty. *Graph Theory with Applications*. North-Holland, 1976.

[Chv83]   Vašek Chvátal. *Linear Programming*. W.H. Freeman and Company, 1983.

[CV89]   James Calvert and William Voxman. *Linear Programming*. Harcourt Brace Jovanovich, 1989.

[Dan51]   G.B. Dantzig. A proof of the equivalence of the programming problem and the game problem. In T.C. Koopmans, editor, *Activity Analysis of Production and Allocation*, number 13 in Cowles Commission Monograph, pages 330–338. John Wiley & Sons, 1951.

[Dre81]   Melvin Dresher. *The Mathematics of Games of Strategy, Theory and Applications*. Dover Publications, 1981.

[Eps67]   Richard A. Epstein. *The Theory of Gambling and Statistical Logic*. Academic Press, 1967.

[Gas86]   Saul I. Gass. *Linear Programming, Methods and Applications*. McGraw-Hill, 1986.

[Jon80]   A.J. Jones. *Game Theory: Mathematical Models of Conflict*. Ellis Horwood Limited, Chichester, 1980.

[KM75]   Donald E. Knuth and Ronald W. Moore. An analysis of alpha-beta pruning. *Artificial Intelligence*, 6:293–326, 1975.

[Lev84]   David Levy. *The Joy of Computer Chess*. Prentice-Hall, Inc., 1984.

[Lev88]   David Levy, editor. *Computer Games I*. Springer-Verlag, 1988.

[LR57]   R. Duncan Luce and Howard Raiffa. *Games and Decisions: Introduction and Critical Survey*. John Wiley & Sons, 1957.

[Luc68]   W.F. Lucas. A game with no solution. *Bulletin of the American Mathematical Society*, 74:237–239, 1968.

[Lue84]   David G. Luenberger. *Linear and Nonlinear Programming*. Addison-Wesley, 1984.

[McK52]   J.C.C. McKinsey. *Introduction to the Theory of Games*. McGraw-Hill, 1952.

[MMS83]   Albert H. Morehead and Geoffrey Mott-Smith. *Hoyle's Rules of Games*. Signet, 1983.

[Mor68]   Oskar Morgenstern. The theory of games. In *Mathematics in the Modern World*, chapter 39, pages 300–303. W. H. Freeman and Company, 1968. First appeared in *Scientific American*, May, 1949.

[Mur52]   H.J.R. Murray. *A History of Board-Games Other Than Chess*. Oxford University Press, 1952. Reprinted by Hacker Art Books, New York, 1978.

[Nas50]   J.F. Nash. The bargaining problem. *Econometrica*, 18:155–162, 1950.

[Nas51]   J.F. Nash. Non-cooperative games. *Annals of Mathematics*, 54(2):286–295, 1951.

[NT93]   Evar D. Nering and Albert W. Tucker. *Linear Programs and Related Problems*. Academic Press, 1993.

[Owe82]   Guillermo Owen. *Game Theory*. Academic Press, 1982.

[PM91]   M.H. Protter and C.B. Morrey. *A First Course in Real Analysis*. Springer-Verlag, second edition, 1991.

[Pou92]   William Poundstone. *Prisoner's Dilemma*. Doubleday, 1992.

[Rap66]   Anatol Rapoport. *Two-Person Game Theory: The Essential Ideas*. The University of Michigan Press, 1966.

[Rap68]   Anatol Rapoport. The use and misuse of game theory. In *Mathematics in the Modern World*, chapter 40, pages 304–312. W.H. Freeman and Company, 1968. First appeared in *Scientific American*, December, 1962.

[Rap70]   Anatol Rapoport. *N-Person Game Theory*. The University of Michigan Press, 1970.

[Rus84]   Laurence Russ. *Mancala Games*. The Folk Games Series, no. 1. Reference Publications, Algonac, Michigan, 1984.

[Sag93]   Carl Sagan. A new way to think about rules to live by. *Parade Magazine*, pages 12–14, November, 1993.

[Sha53]   L.S. Shapley. A value for *n*-person games. In *Contributions to the Theory of Games, II*, Ann. Math. Studies No. 28, pages 307–317. Princeton University Press, 1953.

[Sol84]   Eric Solomon. *Games Programming*. Cambridge University Press, 1984.

[Str89]   James K. Strayer. *Linear Programming and Its Applications*. Springer-Verlag, 1989.

[Tho66]   Edward O. Thorp. *Beat the Dealer: A Winning Strategy For the Game of Twenty One*. Vintage Books, 1966.

[Tho84]   L.C. Thomas. *Games, Theory and Applications*. Ellis Horwood Limited, Chichester, 1984.

[Val64]   Frederick A. Valentine. *Convex Sets*. McGraw-Hill, 1964.

[vNM44]   John von Neumann and Oskar Morgenstern. *Theory of Games and Economic Behavior*. Princeton University Press, 1944.

[Vor77]   N.N. Vorob'ev. *Game Theory, Lectures For Economists and Systems Scientists*. Springer-Verlag, 1977. Translated and supplemented by S. Kotz.

[Wan88]   Wang Jianhua. *The Theory of Games*. Tsinghua University Press, Beijing and Clarendon Press, Oxford, 1988.

[Wil66]   J.D. Williams. *The Compleat Strategyst*. McGraw-Hill, 1966.

Index

Universitext   *(continued)*

Meyer: Essential Mathematics for Applied Fields
Mines/Richman/Ruitenburg: A Course in Constructive Algebra
Moise: Introductory Problems Course in Analysis and Topology
Morris: Introduction to Game Theory
Polster: A Geometrical Picture Book
Porter/Woods: Extensions and Absolutes of Hausdorff Spaces
Radjavi/Rosenthal: Simultaneous Triangularization
Ramsay/Richtmyer: Introduction to Hyperbolic Geometry
Reisel: Elementary Theory of Metric Spaces
Rickart: Natural Function Algebras
Rotman: Galois Theory
Rubel/Colliander: Entire and Meromorphic Functions
Sagan: Space-Filling Curves
Samelson: Notes on Lie Algebras
Schiff: Normal Families
Shapiro: Composition Operators and Classical Function Theory
Simonnet: Measures and Probability
Smith: Power Series From a Computational Point of View
Smoryski: Self-Reference and Modal Logic
Stillwell: Geometry of Surfaces
Stroock: An Introduction to the Theory of Large Deviations
Sunder: An Invitation to von Neumann Algebras
Tondeur: Foliations on Riemannian Manifolds
Wong: Weyl Transforms
Zhang: Matrix Theory: Basic Results and Techniques
Zong: Sphere Packings
Zong: Strange Phenomena in Convex and Discrete Geometry